*Fixation, dehydration
and embedding of
biological specimens*

Practical Methods in
ELECTRON MICROSCOPY

Edited by

AUDREY M. GLAUERT

Strangeways Research Laboratory

Cambridge

1975

NORTH-HOLLAND PUBLISHING COMPANY – AMSTERDAM · OXFORD

AMERICAN ELSEVIER PUBLISHING CO., INC. – NEW YORK

FIXATION, DEHYDRATION AND EMBEDDING OF BIOLOGICAL SPECIMENS

AUDREY M. GLAUERT

Strangeways Research Laboratory

Cambridge, England

1975

NORTH-HOLLAND PUBLISHING COMPANY – AMSTERDAM · OXFORD

AMERICAN ELSEVIER PUBLISHING CO., INC. – NEW YORK

Library of Congress Catalog Card Number 74–76062
North-Holland ISBN for the series 0 7204 4250 8
North-Holland ISBN for this book 0 7204 4257 5
American Elsevier ISBN 0 444 106669

Publishers:

NORTH-HOLLAND PUBLISHING COMPANY – AMSTERDAM
NORTH-HOLLAND PUBLISHING COMPANY, LTD. – OXFORD

Sole Distributors for the U.S.A. and Canada:
AMERICAN ELSEVIER PUBLISHING COMPANY, INC.
52 VANDERBILT AVENUE
NEW YORK, N.Y. 10017

This book is the laboratory edition of Volume 3, Part I, of the series 'Practical Methods in Electron Microscopy'.
Volume 3 of the series contains the following parts:
Part I Fixation, dehydration and embedding of
        biological specimens
        by Audrey M. Glauert
Part II Ultramicrotomy
        by Norma Reid

Titles of earlier volumes published in this series:
Volume 1 Part I Specimen preparation in materials science
                  by P. J. Goodhew
          Part II Electron diffraction and optical diffraction techniques
                  by B. E. P. Beeston, R. W. Horne, R. Markham
Volume 2 Principles and practice of electron microscope operation
          by A. W. Agar, R. H. Alderson, D. Chescoe

Transferred to digital printing 2005
Printed and bound by Antony Rowe Ltd, Eastbourne

Contents

Acknowledgements . . . . *IX*

Chapter 1. Introduction *1*

1.1 Electron microscopy of biological specimens; the need for ultrathin sections 1
1.2 The aim of the procedure 2
1.3 Criteria of good preservation 2

Chapter 2. Fixatives *5*

2.1 The development of fixatives for electron microscopy 6
2.2 Vehicles for fixatives 7
    2.2.1 Choice of vehicle 8
    2.2.2 pH of buffers 8
    2.2.3 Osmolarity 9
    2.2.4 Ionic constitution 10
    2.2.5 Phosphate buffers 12
    2.2.6 Cacodylate buffers 16
    2.2.7 Veronal–acetate buffers 18
    2.2.8 Collidine buffers 21
2.3 Osmium tetroxide fixatives 23
    2.3.1 Chemical and morphological effects of osmium tetroxide fixatives. . 24
    2.3.2 Osmium tetroxide solutions 25
    2.3.3 Phosphate buffered osmium tetroxide fixatives 26
    2.3.4 Cacodylate buffered osmium tetroxide fixatives 27
    2.3.5 Veronal–acetate buffered osmium tetroxide fixatives 28
    2.3.6 Collidine buffered osmium tetroxide fixatives 30
2.4 Glutaraldehyde fixatives 30
    2.4.1 Chemical and morphological effects of glutaraldehyde fixatives . . 31
    2.4.2 Glutaraldehyde solutions 38
    2.4.3 Preparation of glutaraldehyde fixatives 41

2.5   Formaldehyde fixatives 44
      2.5.1   Chemical and morphological effects of formaldehyde fixatives . . . 44
      2.5.2   Formaldehyde solutions 45
      2.5.3   Preparation of paraformaldehyde fixatives 45
      2.5.4   Paraformaldehyde fixatives for very large tissue blocks 46
      2.5.5   Paraformaldehyde–glutaraldehyde fixatives 47
2.6   Acrolein and other aldehyde fixatives 48
      2.6.1   Acrolein fixatives 48
      2.6.2   Acrolein–glutaraldehyde fixatives 49
      2.6.3   Acrolein–glutaraldehyde–paraformaldehyde fixatives 50
      2.6.4   Other aldehyde fixatives 50
2.7   Permanganate fixatives 50
      2.7.1   Potassium permanganate fixatives 50
      2.7.2   Sodium permanganate fixatives 53
      2.7.3   Lanthanum permanganate fixatives 55
      2.7.4   Other permanganate fixatives 57
2.8   Mixed fixatives 57
      2.8.1   Osmium tetroxide–potassium dichromate fixatives 57
      2.8.2   Glutaraldehyde–osmium tetroxide fixatives 58
      2.8.3   Fixatives containing potassium ferricyanide 60
      2.8.4   Aldehyde fixatives containing trinitro compounds 61
      2.8.5   Fixatives containing digitonin 64
      2.8.6   Other additives for fixatives 64
2.9   Uranyl acetate fixatives 65

Chapter 3.   Fixation methods *73*

3.1   Time and temperature of fixation 75
      3.1.1   Primary fixation 75
      3.1.2   Wash 76
      3.1.3   Secondary fixation 77
3.2   Fixation schedules 77
3.3   Organs and tissues 78
      3.3.1   Immersion fixation 78
      3.3.2   *In vivo* fixation 79
      3.3.3   Perfusion fixation 80
              3.3.3a   Anaesthesia 80
              3.3.3b   The route of injection of perfusion fluids 82
              3.3.3c   Fluids and fixatives for perfusion 83
      3.3.4   Fixation of very large specimens 85
      3.3.5   Organ cultures and colonies growing on agar 86
3.4   Botanical specimens 86
3.5   Monolayers of cells 90
3.6   Isolated cells 91
      3.6.1   Fixation in a pellet 91
      3.6.2   Fixation in suspension in medium 93
      3.6.3   Encapsulating methods for pellets 95
      3.6.4   Encapsulating methods for isolated cells 98
      3.6.5   Collecting cells on Millipore filters 99
3.7   Cell fractions 99
3.8   Methods for very small quantities of material 100

3.9 Special fixation schedules and methods. 102
   3.9.1 Lipids 102
   3.9.2 Cells with thick walls 102
   3.9.3 Bacteria 103
   3.9.4 Marine organisms. 104
   3.9.5 Organisms living in extreme conditions 104
   3.9.6 Special fixation schedules 104

Chapter 4. Dehydration *111*

4.1 Chemical and morphological effects of dehydration 111
4.2 Dehydration schedules 112
   4.2.1 Standard dehydration schedule 113
   4.2.2 Rapid dehydration 113
   4.2.3 Partial dehydration 116
4.3 Dehydrating agents 117
   4.3.1 Ethylene glycol 118
   4.3.2 Polyethylene glycol 119
   4.3.3 Propylene oxide 119
   4.3.4 Other dehydrating agents 120

Chapter 5. Embedding *123*

5.1 Embedding media for electron microscopy 123
   5.1.1 Chemical and morphological effects of embedding media 125
5.2 Standard embedding methods 125
5.3 Epoxy resins 130
   5.3.1 Epoxy resin embedding media 130
   5.3.2 Characteristics of epoxy resin embedding media 136
      5.3.2a Viscosity 136
      5.3.2b Hardness of the final block. 137
      5.3.2c Sectioning properties 141
   5.3.3 Preparation of epoxy resin embedding media 143
   5.3.4 Embedding schedules for epoxy resins 144
      5.3.4a Standard embedding schedule 144
      5.3.4b Embedding at room temperature 147
      5.3.4c Rapid embedding 148
5.4 Polyester resins 148
   5.4.1 Polyester resin embedding media 148
   5.4.2 Preparation of polyester resin embedding media 150
   5.4.3 Embedding schedules for polyester resins 151
      5.4.3a Embedding schedule for Vestopal W 151
      5.4.3b Rapid embedding with Vestopal W 152
      5.4.3c Embedding schedule for Rigolac 152
      5.4.3d Embedding schedule for styrene/Rigolac 153
5.5 Methacrylates 153
   5.5.1 Methacrylate embedding media 153
   5.5.2 Preparation of methacrylate embedding media 154
   5.5.3 Embedding schedules for methacrylates 155

5.6   Water-soluble embedding media.   156
      5.6.1   Water-soluble epoxy resins   156
              5.6.1a   Dehydration and embedding schedule for Durcupan   . . .   158
              5.6.1b   Dehydration in Durcupan followed by conventional embedding   158
      5.6.2   Water-soluble methacrylates.   159
              5.6.2a   Glycol methacrylate.   159
              5.6.2b   Hydroxypropyl methacrylate   160
              5.6.2c   Embedding in the presence of water   161
5.7   Other embedding media   161
      5.7.1   Gelatin   161
      5.7.2   Urea-aldehyde.   162
      5.7.3   Protein-aldehyde   162
      5.7.4   Polyampholytes   162
5.8   Special embedding methods   163
      5.8.1   Flat embedding   163
      5.8.2   Monolayers of cells   164
              5.8.2a   Removal of cells from the substrate before embedding   165
              5.8.2b   Separation of hardened blocks from the substrate; 'inverted
                       capsule' technique   165
              5.8.2c   Separation of hardened blocks from the substrate; flat embedding
                       technique.   167
              5.8.2d   Cells not separated from the substrate   167
      5.8.3   Embedding methods for very small quantities of material   170

Chapter 6.   *Low-temperature methods* *177*

6.1   Fixation, dehydration and embedding at low temperatures   177
6.2   Freeze-drying   178
6.3   Freeze-substitution   178
6.4   Cryo-ultramicrotomy   185

Appendix   *Commercial suppliers of equipment and materials for fixation,*
             dehydration and embedding *187*

Index for list of suppliers. *199*

Subject index *201*

Acknowledgements

I am extremely grateful for the valuable help and constructive criticism I received from Dr. J. R. Gibbins, Dr. P. R. Lewis, Dr G. Millonig, Mr. R. A. Parker and Mrs. M. E. Van Steveninck during the preparation of this book. In particular I wish to thank Dr. Lewis for contributing the section on perfusion fixation, and Mrs. Van Steveninck for the section on the fixation of botanical specimens in Chapter 3.

I am also indebted to the following for permission to reproduce micrographs and figures: Dr. T. C. Appleton, Dr. J. G. Bluemink, Dr. M. W. Brightman, Dr. S. Bullivant, Dr. S. Busson-Mabillot, Dr. F. Carson, Dr. E. Dimmock, Dr. C. Franzini-Armstrong, Dr. J. Gil, Dr. P. V. Johnston, Dr. H. Kushida, Dr. S. Malamed, Professor R. G. Monroe, Dr. B. I. Roots, Dr. J. McD. Tormey and Professor E. R. Weibel.

Finally I would like to thank Mr. R. A. Parker for assistance with the illustrations and for providing Figures 3.2, 5.4, 5.5 and 5.6, Mrs. Doreen Stebbings for typing the manuscript and my husband for his constant encouragement and patience.

November 1973   AUDREY M. GLAUERT

Chapter 1

Introduction

One of the major consequences of the development of techniques for the fixation, embedding and sectioning of specimens for electron microscopy has been a complete reappraisal of the micro-anatomy of biological tissues and organisms during the past 25 years. The fact that electron micrographs of ultrathin sections are used to illustrate practically every textbook and monograph, and in many research papers, in cell biology and anatomy is sufficient indication of the importance of the technique as one of the basic methods of modern biology and medical science.

This book describes in detail the methods for the fixation, dehydration and embedding of biological specimens, while techniques for preparing ultrathin sections (Reid 1974) and for staining sections to provide contrast in the electron microscope (Lewis et al. 1974) are described in companion volumes. In common with other books in this series the aim has been to describe the techniques of electron microscopy in sufficient detail to enable the isolated worker to carry them out successfully. No specialist knowledge is assumed. Only well-established techniques are discussed in full, although many others are outlined briefly and reference given to the original papers in which they are described. The literature survey for this book ended in January 1973.

1.1 Electron microscopy of biological specimens; the need for ultrathin sections

It is necessary to prepare ultrathin sections of biological specimens because of the limited penetration of the electron beam in the electron microscope operating at conventional accelerating voltages up to 100 kV. For adequate penetration and resolution the specimen should not be thicker than 100 nm, and preferably considerably thinner. Only a few specimens, such as virus particles and isolated sub-cellular organelles, are thin enough to be viewed directly, and consequently considerable attention has been paid to the development of fixation, embedding and sectioning techniques.

Ultrathin sections are not suitable for the study of fine structure at the optimum resolution of the electron microscope (0.5 nm for most standard instruments) since structural detail smaller than about 2 nm does not appear to be preserved by the fixation and embedding procedures at present available. Fortunately biological specimens possess a great wealth of structural detail in the 2 to 100 nm range and here the study of sections provides information that can be obtained in no other way.

There is a well-defined series of steps in the fixation, dehydration and embedding of biological specimens and in general it is only necessary to adopt a very standard routine which is now followed in many laboratories. Each of the steps of the preparative procedure is described in detail in the following chapters, together with the modifications of the basic methods that are required for special types of specimen.

1.2   The aim of the procedure

The final aim of the whole procedure is to produce blocks which can be sectioned without undue difficulty and which contain specimens in which the fine structure is preserved with as little alteration as possible when compared with the living organism. As shown by Reid (1974) the production of suffi- ciently thin sections is usually no problem with standard embedding media, but the problem of how to preserve the specimen in a life-like state is by no means solved. At each stage of the procedure, from the initial fixation onwards, changes are inevitably introduced; material is extracted, dimen- sions are altered (Fig. 1.1) and molecular rearrangement occurs. The best that can be done is to keep these changes to a minimum and the achievement of this is the over-riding consideration in the choice of preparative procedures.

1.3   Criteria of good preservation

Since some structures observed in electron micrographs of ultrathin sections cannot be visualised by any other method, there is an inevitable element of subjectivity in judging the quality of preservation achieved. There are a number of criteria, however, which can be applied:

(1) The final appearance of the specimen should be similar to that of

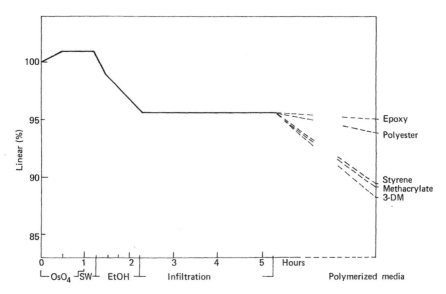

Fig. 1.1. An example of dimensional changes occurring during the preparative procedure. The graph shows the percentage changes in the diameters of sea urchin eggs during fixation in osmium tetroxide (OsO_4), washing in sea water (SW), dehydration in ethanol (EtOH), and infiltration and embedding in five different embedding media. 3-DM is three-dimensional methacrylate (from Kushida 1962).

the living specimen viewed with the light microscope. It is rarely profitable to examine a specimen in the electron microscope before a full understanding of its appearance in the light microscope has been obtained. The examination of 'thick' (0.5 to 2.0 μm) sections of blocks prepared for electron microscopy (Reid 1974) has recently become a powerful tool in cytology.

(2) Obvious structural damage should be absent. Membranes should be continuous; the space between the two nuclear membranes (the peri-nuclear space) should not be swollen and empty; there should be no empty spaces in the ground substance or within organelles.

(3) Loss of material and dimensional changes should be kept to a minimum. If the specimen contains a regular periodic structure, such as a myofibril, this can be used as an internal standard to follow dimensional changes.

(4) A structure should have the same general appearance after fixation with a number of different fixatives.

Different specimens require different preparative procedures, particularly at the fixation stage (Chapter 3) and the beginner will find that it is always worthwhile to first follow methods already used successfully by others for similar specimens, and then to introduce modifications if these prove to be necessary.

REFERENCES

Kushida, H. (1962), A study of cellular swelling and shrinkage during fixation, dehydration and embedding in various standard media, J. Electron Microscopy *11*, 135.
Lewis, P. R., D. P. Knight and M. A. Williams (1974), Staining methods for thin sections, in: Practical methods in electron microscopy, A. M. Glauert, ed. (North-Holland, Amsterdam).
Reid, N. (1974), Ultramicrotomy, in: Practical methods in electron microscopy, A. M. Glauert, ed. (North-Holland, Amsterdam).

Chapter 2

Fixatives

Fixation is the first step in the preparation of biological specimens for examination by electron microscopy of thin sections and has the aim of 'stabilising cellular organisation to such an extent that ultrastructural relations are preserved despite the subsequent rather drastic treatments of dehydration, embedding and exposure to the electron beam' (Riemersma 1968). Inevitably, however, the structure of the living specimen is modified during this stabilisation procedure, and these modifications may well be greatest for those fixatives that produce the best protection against the effects of the subsequent steps in the preparative procedure. With some fixatives obvious disruption of fine structure is observed; discontinuities are present in membranes, tubules are transformed into chains of vesicles and, with certain fixatives such as permanganates, cell components (ribosomes, myofilaments, cytoplasmic fibrils) are completely destroyed. Some of these effects depend as much on the nature of the specimen as on the fixative since they result from interactions between the two, and in consequence a number of conflicting criteria influence the choice of a fixative for a particular specimen.

In extreme examples the fixative may induce noticeable physiological changes, such as the contraction or relaxation of muscle fibres or the triggering of secretion by solutions containing calcium. The effects of individual fixatives are described in the separate sections below, but it is important to realise from the outset that there is no one perfect fixative for all types of specimen; not only do different tissues require different fixatives, but it may even be necessary to use different fixatives at different stages of development of a tissue (e.g. Kelly and Zacks 1969) or for different structural elements within a tissue (e.g. Doggenweiler and Heuser 1967). The general aim is usually to obtain satisfactory preservation of the tissue or organ as a whole, and not to achieve the best possible preservation of a given cell type with disregard to the fate of others (Palay et al. 1962). For some studies, however, it will be necessary to choose the conditions so as to obtain the optimum fixation for certain structures even if this is to the detriment of others.

The composition and properties of the various fixatives currently used in electron microscopy are described in this chapter, while suggestions on the choice of a fixative for a particular study are made in the following chapter (Chapter 3) which is concerned with methods of fixation.

2.1   The development of fixatives for electron microscopy

A fixative consists of a fixing agent (e.g. osmium tetroxide) in a suitable vehicle (usually a buffer plus various salts, § 2.2).

Most of the fixatives developed for light microscopy are inadequate for electron microscopy since much of the very fine detail visible by electron microscopy is not preserved, and extensive extraction and precipitation occurs (Baker 1965). Osmium tetroxide is the best of the light microscope fixatives for preserving fine structure and it was the basis of the earliest fixatives specially developed for electron microscopy (Palade 1952; Rhodin 1954; Zetterqvist 1956).

The adequate fixation of botanical tissues was less easily obtained with osmium tetroxide (Mollenhauer 1959; Afzelius 1962) and consequently potassium permanganate (Luft 1956) became popular as an alternative fixative both for plant tissues and for membranes (e.g. Robertson 1959).

A major change in fixation procedures followed the publication in 1963 by Sabatini, Bensch and Barrnett of the results of their search for fixatives suitable for electron microscope cytochemistry. They recommended the use of aldehydes, particularly glutaraldehyde, as primary fixatives, followed by a secondary fixation with osmium tetroxide. These studies made a major contribution to the development of new techniques in cytochemistry (see Lewis et al. 1974), but the importance of the work in the present context is that it drew attention to the value of aldehydes as fixatives for electron microscopy. Other aldehydes had been used previously, but not glutaraldehyde, and it was this aldehyde which was found to give excellent preservation of a wide range of animal and plant tissues when followed by a second fixation with osmium tetroxide. The procedure has the advantage that most specimens can be stored in buffer after glutaraldehyde fixation for considerable periods of time before further processing. The convenience of this double fixation and the good preservation of fine structure obtained has made it the standard procedure in most laboratories. The success of glutaraldehyde led to a re-examination of other aldehydes as primary fixatives, with the result that it was found that formaldehyde, freshly prepared from paraformaldehyde (Robertson et al. 1963) compares favourably with glutaraldehyde and has the added advantage of more rapid penetration into the specimen. Subsequently various mixtures of aldehydes have been tested and some of them, particularly a mixture of paraformaldehyde and glutaraldehyde (Karnovsky 1965), have proved to be superior to either aldehyde used alone for many specimens.

Recent years have seen the further developments of fixatives for electron microscopy, including the use of uranyl acetate as a third fixative, following glutaraldehyde and osmium tetroxide. When uranyl acetate is used in this way, before dehydration and embedding, it is often described as a 'block stain' and its properties as such are considered by Lewis et al. (1974). However, under these conditions uranyl acetate also acts as a fixative, particularly for lipid components, and so it will be included here also.

All these various fixatives and their properties are described in the sections below. In general, when starting a study of a particular specimen it is advisable to first follow the most recent methods used successfully by others (see Chapter 3) and then to make any modifications required.

2.2 Vehicles for fixatives

An ideal fixative would exactly match the natural environment of a living tissue with respect to pH, osmolarity and ionic constitution. For isolated cells (see § 3.6.2) an approximate match is obtained by dissolving the fixing agent in the growth medium or by exposing cells suspended in their growth medium to the vapours of the fixative, but problems arise when fixing uni- or multi-cellular organisms which live in environments of very low or very high ionic strength, or at alkaline or acid pH (§ 3.9.5), and when fixing organs and tissues. The best choice of components for the vehicle of the fixative is usually a matter of trial and error since there is little detailed knowledge of the constituents of the fluids and matrices surrounding cells or of the intracellular composition of the cells in various tissues. In addition, the fixing agent itself may have unpredictable effects on the permeability of membranes and therefore of cells and organelles (§ 2.2.3). Fortunately when a piece of tissue is immersed in a fixative the strong buffering capacity of the tissue proteins tends to keep the pH within a narrow range (Millonig and Marinozzi 1968). For this reason good fixation of some tissues can be obtained with a number of different fixatives, or even with unbuffered fixatives (Claude 1962; Malhotra 1962). Other tissues, such as the epithelia of the kidney (Rhodin 1954) and intestine (Zetterqvist 1956), and isolated cells, are much more sensitive to the composition of the fixative.

2.2.1 CHOICE OF VEHICLE

The properties required of a vehicle for a fixative are:

(i) Ability to maintain a constant pH during fixation

(ii) Suitable osmolarity (when mixed with the fixing agent) 'so that cells and organelles neither swell nor shrink during fixation

(iii) Suitable ionic constitution so that materials are neither extracted or precipitated during fixation.

Additional criteria, such as lack of toxicity and colloid osmotic pressure, govern the choice of vehicles for fixation by perfusion (see § 3.3.3).

The buffers commonly used for electron microscopy include phosphate, cacodylate, veronal-acetate and collidine, the majority of fixatives being buffered with phosphate or cacodylate. Veronal-acetate and collidine buffers will be included for completeness and reasons given why they have largely been superseded as buffers for primary fixatives. They are suitable as buffers for secondary fixatives but they suffer from a number of practical disadvantages and are best avoided, if possible. In fact it does not appear to be generally realised that all the buffers used in electron microscopy, with the notable exception of phosphate buffers, are potentially harmful, both to the specimen and to the microscopist, since they contain toxic components such as barbiturates (veronal–acetate) and arsenical compounds (cacodylate). They also have an unpleasant smell, which is worst for collidine. Great care should be taken not to inhale the vapours from these buffers; they should always be handled in a fume cupboard and should be flushed away with large quantities of water. It is strongly recommended that phosphate buffers be used whenever possible, particularly for routine work.

2.2.2 pH OF BUFFERS

Fixatives for electron microscopy are usually adjusted to a pH near neutrality using the chosen buffer system. For the majority of tissues there is little evidence that the pH of the fixative is at all critical within the range pH 6.5 to 8.0, although from model experiments with albumin, Millonig and Marinozzi (1968) suggest that phosphate buffered osmium tetroxide fixatives should be adjusted to a pH of 7.0 or below. Slightly acidic (pH 6.0) osmium tetroxide fixatives have been recommended for the preservation of nuclear material and spindle fibres. With the advent of aldehyde fixatives, primary fixation with glutaraldehyde at approximately neutral pH is recommended for all specimens.

2.2.3 OSMOLARITY

It is not possible to calculate the correct osmolarity for a fixative since insufficient is known about the internal osmotic pressure of living cells. In addition, it is clear that alterations in the permeability of membranes occur during fixation. After osmium tetroxide fixation membranes become freely permeable to macromolecules such as ferritin (Tormey 1965), while after glutaraldehyde or formaldehyde fixation the permeability properties of membranes are only partially preserved (Jard et al. 1966). Although there are rapid changes in the osmotic properties of cell envelopes at fixation, there is no loss of the intracellular colloid-osmotic equilibrium. Therefore fixing and rinsing solutions have to be adjusted to this equilibrium which, in terms of osmotic pressure, may differ from the iso-osmotic environment of the cell (Millonig and Marinozzi 1968).

In these circumstances the only practical course is to select a fixative following procedures used in earlier studies on a similar tissue and then to make adjustments if signs of swelling or shrinkage are observed. If a number of different cell types is present it may be necessary to use slightly different fixatives for the optimum preservation of each cell type, since there is evidence that the osmotic pressure is different in different cells and is generally higher than that of blood plasma (Palade 1956). In general a hypertonic fixative is less damaging than a hypotonic one.

It is sometimes necessary to record the osmolarity of a fixative when studying osmotically sensitive specimens so that the fixative may be standardised from one experiment to another. Workers for whom osmolarity is important, such as those studying isolated cells, favour the estimation of osmolarity from the measurement of freezing-point depression with an osmometer, since calculations from formulae do not take into account the degree of dissociation of the various salts (Millonig 1962; Maser et al. 1967).

The osmolarity of the fixative is adjusted by altering the concentration of the buffer (and not the concentration of the fixing agent) or by the addition of sodium chloride or non-ionic compounds, such as glucose or sucrose. Glucose and sucrose should not be used in primary osmium tetroxide fixatives since they inhibit proper 'fixation' of albumin in model experiments (Millonig 1964; Millonig and Marinozzi 1968), and may lead to an increased extraction of soluble material during fixation. They are suitable, however, for secondary osmium tetroxide fixatives and for aldehyde fixatives. Monovalent ions are preferred to divalent ions for the adjustment of osmolarity since divalent ions are more likely to produce specific ion effects and may induce a granular precipitation of proteins (Millonig and Marinozzi 1968). Sodium chloride is usually used and should be added to the buffer before the pH is finally adjusted since it may alter the pH. Glutaraldehyde and formaldehyde are not themselves osmotically active (§§ 2.4.1 and 2.5.1) and it has been suggested that their contribution to the osmolarity of the fixative can be ignored (Bone and Ryan 1972).

2.2.4   IONIC CONSTITUTION

When the composition of the fixative differs from that of the natural extracellular fluids of the tissue there is a likelihood of extraction of cellular materials and/or deposition of fixative components. These effects occur in the period before the tissue is adequately 'fixed' and are particularly marked with fixatives in which the fixing agent penetrates (e.g. osmium tetroxide) or reacts (e.g. formaldehyde) slowly. Consequently when osmium tetroxide or formaldehyde is used as the primary fixative the ionic constitution of the vehicle is important (Wood and Luft 1965; Carson et al. 1972) and care should be taken to match it as closely as possible to the tissue fluids. Also non-ionic additives should be used to adjust the osmolarity of the fixative if there is any evidence that the addition of monovalent ions will have any effect.

It seems likely that the superiority of glutaraldehyde over osmium tetroxide as a primary fixative, particularly for non-membraneous components of the cytoplasm (Figs. 2.1a and b) and extracellular materials, is partly due to the faster rate of reaction of glutaraldehyde with the result that fixation occurs before the extraction of cell components by the buffer is noticeable. This would explain why the choice of vehicle is important in the fixation of tissues with osmium tetroxide (Trump and Ericsson 1965), but not with glutaraldehyde (Gil and Weibel 1968) and why no differences are found in the results with differently buffered glutaraldehyde fixatives so long as the buffers have approximately the same osmolarity (Maunsbach 1966b; Busson-Mabillot 1971).

It is sometimes possible to relate these extractive effects directly to the presence or absence of specific ions. For example, bicarbonate ions have a specific destructive effect on some microtubules (Schultz and Case 1968)

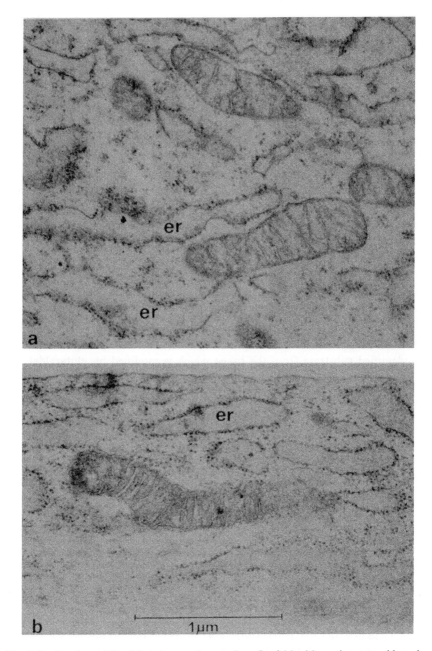

Fig. 2.1. Rat dermal fibroblasts in monolayer culture fixed (a) with osmium tetroxide and (b) with glutaraldehyde, followed by osmium tetroxide. The double fixation gives much better preservation of the contents of the cisternae of the endoplasmic reticulum (er) and of cytoplasmic filaments. (Unpublished micrographs from a study by Audrey M. Glauert and Mary R. Daniel.)

while divalent cations are necessary for the preservation of membrane lipids (e.g. Mitchell 1969). The presence of divalent cations in fixatives is not wholly beneficial since this may induce a precipitation of proteins and give a granular appearance to cell structures (Millonig and Marinozzi 1968). Magnesium is preferred to calcium by Millonig (1973) because it is a smaller molecule and, at reasonable concentrations, does not precipitate proteins. In addition calcium may exert significant physiological effects, such as the stimulation of muscle fibres or the triggering of secretion.

Artefacts due to precipitation are less easy to detect than those due to extraction since such precipitates are unlikely to be noticed unless they are comparatively large and dense. A striking and extreme example of such precipitates is provided by the observation of Gil and Weibel (1968) that spherical dense granules appear on the alveolar surfaces of rat lungs and on erythrocytes after fixation with glutaraldehyde followed by osmium tetroxide in phosphate buffer (Fig. 2.2). The appearance of the granules is independent of the buffer used with the glutaraldehyde or for washing, and the granules are only seen when the osmium tetroxide is buffered with phosphate and not with other buffers. They also do not appear after primary fixation with phosphate buffered osmium tetroxide, and Gil and Weibel (1968) concluded that the precipitates are a complex containing osmium, phosphate and an unidentified substrate from the tissue which is modified by glutaraldehyde fixation.

In general, the possibility of the formation of precipitates during fixation should be borne in mind, particularly when using fixatives containing divalent cations or phosphates.

2.2.5 PHOSPHATE BUFFERS

Phosphate buffers have the advantage that they are the most 'physiological' of the buffers used in electron microscopy since they mimic certain components of extracellular fluids and are non-toxic to cells in culture. In consequence they are excellent buffers for fixation by perfusion and for slowly penetrating and reacting fixatives, such as osmium tetroxide. Their only disadvantages are that precipitates are more likely to occur during fixation than with other buffers (see § 2.2.4), and that they become slowly contaminated with micro-organisms.

A number of different phosphate buffers have been used and there is little evidence that one is basically superior to another so long as the osmolarity is the same. Most of the buffers used are based on Sörensen's buffer (see Dawson et al. 1969) which is a mixture of monobasic and dibasic

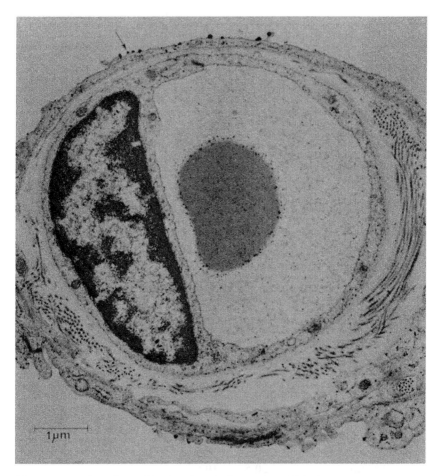

Fig. 2.2. Dog lung fixed with phosphate buffered glutaraldehyde, washed in phosphate buffer, and post-fixed with phosphate buffered osmium tetroxide. Dense granules are present on the alveolar surface (arrow) and on the surface of an erythrocyte in a capillary. (Unpublished micrograph from a study by Joan Gil and Ewald R. Weibel.)

sodium phosphates. The method of preparation of a 0.1 M phosphate buffer by the method of Sörensen is therefore described first, and then the formulae of the buffers that have been widely used in electron microscopy are given. A phosphate buffer of any desired pH and osmolarity can be prepared following the same basic principles. The pH of phosphate buffers changes little with temperature; the buffers are non-toxic and are stable for several weeks at 4 °C if sucrose or glucose have not been added. Sediments may appear during storage and so it is advisable to make the buffers up freshly each month from the stock solutions.

Method of preparation of phosphate buffer I
(Sörensen, see Dawson et al. 1969)

1. Prepare a 0.2 M solution of dibasic sodium phosphate with

	$Na_2HPO_4 \cdot 2H_2O$	35.61 g
or	$Na_2HPO_4 \cdot 7H_2O$	53.65 g
or	$Na_2HPO_4 \cdot 12H_2O$	71.64 g
	distilled water to make	1000 ml

2. Prepare a 0.2 M solution of monobasic sodium phosphate with

	$NaH_2PO_4 \cdot H_2O$	27.6 g
or	$NaH_2PO_4 \cdot 2H_2O$	31.21 g
	distilled water to make	1000 ml

3. Prepare the 0.1 M phosphate buffer by mixing x ml of 0.2 M dibasic sodium phosphate with y ml of 0.2 M monobasic sodium phosphate and diluting to 100 ml with distilled water.

pH (at 25 °C)	x ml	y ml
5.8	4.0	46.0
6.0	6.15	43.85
6.2	9.25	40.75
6.4	13.25	36.75
6.6	18.75	31.25
6.8	24.5	25.5
7.0	30.5	19.5
7.2	36.0	14.0
7.4	40.5	9.5
7.6	43.5	6.5
7.8	45.75	4.25
8.0	47.35	2.65

The osmolarity of the buffer is adjusted by varying the molarity of the phosphates, or by the addition of sucrose, glucose or sodium chloride. The osmolarity of 0.1 M Sörensen's buffer at pH 7.2 is 226 mosmols (Fahimi and Drochmans 1965b); corresponding values for 0.05 M, 0.075 M and 0.15 M buffers are 118, 180 and 350 mosmols, while addition of 0.18 M sucrose to a 0.1 M buffer raises the osmolarity to 425 mosmols.

Method of preparation of phosphate buffer II
(Maunsbach 1966a)

Prepare the 0.135 M phosphate buffer with

NaH_2PO_4. H_2O 2.98 g
and Na_2HPO_4. $7H_2O$ 30.40 g
distilled water to make   1000   ml

The pH of this buffer is 7.35 and the osmolarity is 298 mosmols. Maunsbach (1966a) used this buffer for fixation of the rat kidney by perfusion with osmium tetroxide, glutaraldehyde or formaldehyde.

Method of preparation of phosphate buffer III
(Karlsson and Schultz 1965)

Prepare the phosphate buffer with

NaH_2PO_4. H_2O 3.31 g
Na_2HPO_4. $7H_2O$ 33.77 g
distilled water to make   1000   ml

The pH of this buffer is 7.4; the osmolarity is 320 mosmols and is equal to that of the cerebro-spinal fluid of rats. Karlsson and Schultz (1965) used this buffer for fixation of the rat central nervous system by perfusion with aldehydes.

Method of preparation of phosphate buffer IV
(Millonig 1964; Karlsson and Schultz 1965)

Prepare the phosphate buffer with

NaH_2PO_4. H_2O 1.8 g
Na_2HPO_4. $7H_2O$ 23.25 g
NaCl   5.0 g
distilled water to make   1000   ml

The pH of this buffer is 7.4. The osmolarity is 440 mosmols and so the buffer is hypertonic to most body fluids. It was recommended by Millonig (1964) for the fixation of very hydrated tissues with osmium tetroxide. The formula of this buffer illustrates how the osmolarity of a phosphate buffer may be adjusted with sodium chloride without altering the molarity of the phosphates. Millonig (1964) suggests an even higher concentration (3%) of sodium chloride for the fixation of marine organisms.

Method of preparation of phosphate buffer V

(Millonig 1961)

1. Prepare the following solutions

   2.26% NaH_2PO_4. H_2O in water   (solution A)

   2.52% NaOH in water   (solution B)

2. Prepare the 0.13 M phosphate buffer with

   solution A   41.5 ml

   solution B   8.5 ml (for pH 7.3).

The pH of this buffer can be adjusted to any required value with solution B without changing the molarity. The buffer is stable for some weeks at 4°C.

2.2.6 CACODYLATE BUFFERS

Cacodylate buffers were first proposed for electron microscopy by Sabatini et al. (1963). They have the advantages that they are easy to prepare, are stable during storage for long periods and do not support the growth of micro-organisms. In addition, they are convenient when a controlled amount of calcium is to be added to the fixative since precipitation does not occur at the low concentrations of calcium (1 to 3 mM) that are usually required. Their main disadvantages are that they contain arsenic, which is toxic and may act as a fixative, and they have an unpleasant smell. They should always be handled in a fume cupboard.

There are few records of the use of cacodylate buffers for primary fixation with osmium tetroxide. When used during primary fixation with aldehydes the quality of preservation is usually similar to that with phosphate buffers, although they give less uniformly good results in the fixation of blood and bone marrow cells with paraformaldehyde (Carson et al. 1972) (Fig. 2.3 and Table 2.1).

Method of preparation of cacodylate buffers

1. Prepare a 0.4 M solution of sodium cacodylate with

   $Na(CH_3)_2AsO_2$. $3H_2O$ 21.4 g

   distilled water to make   250   ml

2. Prepare the 0.2 M cacodylate buffer with

   0.4 M sodium cacodylate   50 ml

   0.2 M HCl   8 ml (approx. for pH 7.2)

   distilled water to make   100 ml

The pH of the buffer is adjusted to the required value with the HCl.

TABLE 2.1

Summary of cytological effects of buffer and temperature variation on paraformaldehyde fixation. (From Carson et al. 1972.)

	s-Collidine	Cacodylate		Phosphate	
	(4°C)	(4°C)	(Room temperature)	(4°C)	(Room temperature)
Membranes	Disrupted	Good	Good	Good	Good
Nucleus					
Chromatin aggregation	None	Peripheral	Peripheral	Peripheral	Peripheral
Density	Low	Moderate	Moderate	Dense	Dense
Perichromatin granules	Present	Present	Present	Present	Present
Peri-nuclear space	Nil	Increased swelling	Nil to slight	Slight	Nil
Mitochondria	Disrupted	Good Some mottled irregularity of matrix	Good	Excellent	Excellent
Endoplasmic reticulum swelling	Disrupted	Slight	Nil	Nil	Nil
Cytoplasmic matrix	Extracted	Varies from dense granular to irregular mottled	Some extraction	Dense uniform	Dense uniform
Preservation of granules					
Neutrophils	Poor	Poor	Fair to good	Excellent	Excellent
Basophils	Poor	Poor	Poor	Poor	Poor
Eosinophils	Fair	Excellent	Good	Excellent	Excellent
Platelets	Poor	Good	Good	Excellent	Excellent

2.2.7   VERONAL–ACETATE BUFFERS

The Michaelis veronal–acetate buffers were chosen for use with osmium tetroxide in the early development of fixatives for electron microscopy in New York (Palade 1952) and Stockholm (Rhodin 1954; Zetterqvist 1956) because they cover the physiological range of pH at constant ionic strength and they appeared to give slightly better results than phosphate buffers (Palade 1956). Later studies have shown that veronal–acetate is a poor buffer

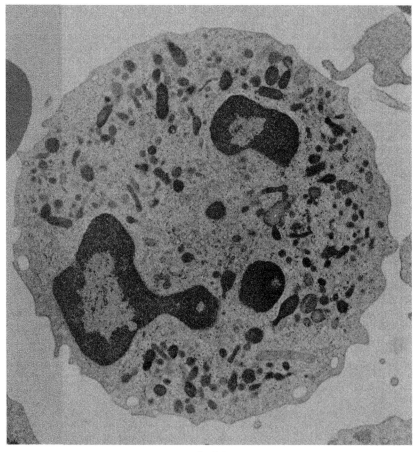

Fig. 2.3a

Fig. 2.3.   Thin sections of neutrophils fixed with paraformaldehyde in various buffers: (a) phosphate buffer; (b) cacodylate buffer; (c) collidine buffer. (Unpublished micrographs from a study by Freida L. Carson.)

system in the region of pH 5.2 to 7.5 (Bennett and Luft 1959) and is useless at physiological pH. It seems likely that the early successes were due to the fact that some tissues are adequately fixed with unbuffered osmium tetroxide (Claude 1962; Malhotra 1962) so that the absence of an effective buffer system was not disastrous. Veronal–acetate buffers are unsuitable for aldehyde fixatives since sodium veronal reacts with aldehydes to produce a substance with no buffering capacity in the physiological range of pH values (Holt and Hicks 1961).

The formulae for the veronal–acetate buffers used in electron microscopy are included here, partly for historical reasons, but also because they are satisfactory, but not recommended, for buffering osmium tetroxide solutions which are to be used as a second fixative, following primary fixation with an aldehyde, and they are also useful for fixation (or block staining) with uranyl acetate since, unlike phosphate and cacodylate buffers, they do not cause the formation of precipitates.

Veronal–acetate buffers contain a barbiturate and so their use should be discouraged except when no alternative is available. They should be handled and disposed of with great care. They cannot be stored for any length of time in the absence of the fixing agent since they support the growth of micro-organisms.

Method of preparation of veronal–acetate buffer I

(Palade 1952)

1. Prepare a veronal–acetate stock solution with

	sodium veronal (barbitone sodium)	2.89 g
	sodium acetate (anhydrous)	1.15 g
or	sodium acetate (hydrated)	1.90 g
	distilled water to make	100 ml

This solution is stable and keeps for some months at 4 °C.

2. Prepare the veronal–acetate buffer with

veronal–acetate stock solution	5.0 ml
distilled water	15.0 ml
0.1 N HCl	5.0 ml (approx.)

The HCl is added gradually until the required pH is reached. This buffer supports the growth of moulds and bacteria and consequently it cannot be stored for any length of time, even at 4 °C.

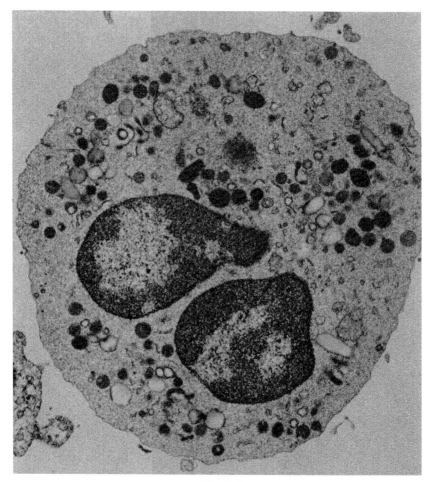

Fig. 2.3b

Method of preparation of veronal-acetate buffer II
(Zetterqvist 1956)

1. Prepare a veronal–acetate stock solution with

sodium veronal (barbitone sodium)	2.94 g
sodium acetate (hydrated)	1.94 g
distilled water to make	100 ml

This solution is stable and keeps for some months at 4 °C.

2. Prepare a Ringer's solution with

sodium chloride	8.05 g
potassium chloride	0.42 g
calcium chloride	0.18 g
distilled water to make	100 ml

3. Prepare the veronal-acetate buffer with

veronal–acetate stock solution	10.0 ml
Ringer's solution	3.4 ml
distilled water	25.0 ml
0.1 HCl	11.0 ml (approx.)

The pH is adjusted to the required value with the HCl. This buffer is not stable and cannot be stored for any length of time.

Method of preparation of veronal-acetate buffer III

(Ryter and Kellenberger 1958)

1. Prepare a veronal–acetate stock solution with

sodium veronal (barbitone sodium)	2.94 g
sodium acetate (hydrated)	1.94 g
sodium chloride	3.40 g
distilled water to make	100 ml

2. Prepare the veronal–acetate buffer with

veronal–acetate stock solution	5.0 ml
distilled water	13.0 ml
1.0 M calcium chloride	0.25 ml
0.1 N HCl	7.0 ml (approx.)

The pH is adjusted to the required value with the HCl. This buffer must be prepared on the day it is to be used since it quickly becomes contaminated with micro-organisms.

2.2.8 COLLIDINE BUFFERS

Bennett and Luft (1959) suggested the use of s-collidine (γ-collidine or 2,4,6-trimethyl-pyridine) as a buffer system for electron microscopy since it exerts its maximum buffering capacity in the neighbourhood of pH 7.4 when half neutralised with a strong acid. In addition the buffer is stable indefinitely at room temperature. Collidine buffers are not suitable, however, for buffering primary osmium tetroxide fixatives since considerable extraction of tissue components occurs during fixation and the subsequent dehydration

(Wood and Luft 1965). They also lead to lysis of the cytoplasmic matrix and extensive destruction of membranes when used to buffer paraformaldehyde fixatives (Carson et al. 1972) (Fig. 2.3c and Table 2.1) and give poorer results for some tissues than phosphate or cacodylate buffers in glutaraldehyde fixatives (Busson-Mabillot 1971) (Fig. 2.8). Collidine buffers are toxic, have a particularly horrid smell and must be handled with extreme care. In

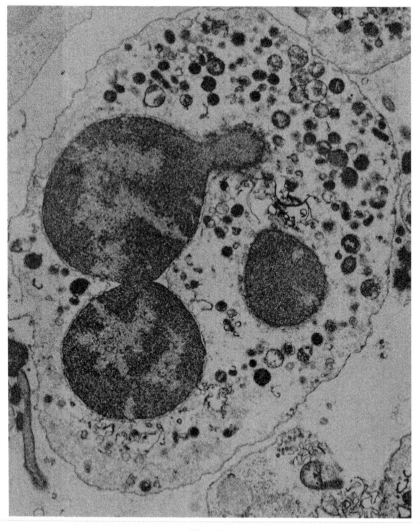

Fig. 2.3c

consequence they cannot be recommended for general use; they are included here because they appear to be adequate for secondary fixation with osmium tetroxide, following an aldehyde (e.g. Gil and Weibel 1968) and are of value for the fixation of very large blocks of tissues where their extractive effects aid the penetration of the fixative (Winborn and Seelig 1970).

It is necessary to use purified *s*-collidine. This is obtainable commercially (see Appendix) although it is expensive. Alternatively it may be purified in the laboratory by distillation (Holt and Hicks 1961).

Method of preparation of collidine buffers

(Bennett and Luft 1959)

1. Prepare a stock solution of *s*-collidine with

   | *s*-collidine (pure) | 2.67 ml |
   | distilled water to make | 50 | ml (approx.) |

2. Prepare the 0.2 M collidine buffer with

s-collidine stock solution	50	ml
1.0 N HCl	9	ml (approx. for pH 7.4)
distilled water to make	100	ml

The pH of the buffer is adjusted with the HCl. It can be stored indefinitely at room temperature.

2.3 Osmium tetroxide fixatives

The value of osmium tetroxide for the preservation of fine detail was apparent as long ago as 1927 when Strangeways and Canti studied the effects of various fixatives on the appearance of living cells by dark-field microscopy. They found that osmium tetroxide was the only fixative of those tested that preserved delicate cytoplasmic processes and caused no alteration in mitochondria and fat droplets. When the first studies on ultra-thin sections were made in the late 1940s and early 1950s it was soon found that osmium tetroxide was preferable for the examination of cellular fine structure to other fixatives developed for light microscopy.

The main disadvantage of osmium tetroxide is that it penetrates and reacts with tissues so slowly that considerable changes in structure can occur before fixation is complete. The ionic constitution of the buffer (see § 2.2.4) has to be chosen carefully since it exerts a profound effect on the fine structure of the specimen (Trump and Ericsson 1965). Also the buffer influences the rate of penetration of the osmium tetroxide. Specimens for fixation must be very small (0.5 to 1.0 mm thick) and even then a central region may still be poorly preserved. In consequence osmium tetroxide has been completely superseded by aldehydes as a primary fixative and the appearance of specimens fixed with osmium tetroxide alone will not be considered in any detail. It still plays an essential role, however, as a second fixative since it reacts with components of tissues, particularly lipids, that are not fixed by aldehydes, and also acts as a stain (Millonig and Marinozzi 1968). During this second fixation the slow rate of penetration of osmium tetroxide is no longer such a disadvantage since the structure of the specimen has already been partially stabilised by the primary aldehyde fixation. Consequently the composition of the buffer is much less important in secondary osmium tetroxide fixatives than in primary fixatives.

2.3.1 CHEMICAL AND MORPHOLOGICAL EFFECTS OF OSMIUM TETROXIDE FIXATIVES

A considerable amount is known about the chemistry of the interactions of osmium tetroxide with isolated lipids, proteins, nucleic acids and carbohydrates and this knowledge is well summarised in reviews by Millonig and Marinozzi (1968), Riemersma (1970) and Hayat (1970). Although the results of such studies are of considerable interest to the molecular biologist, and may well lead to the development of improved fixatives in the future, as yet they are of little practical help to the microscopist in the choice of a fixative, since there remains considerable uncertainty about the way osmium tetroxide interacts with cellular components *in situ* (Riemersma 1970).

Studies of the effects produced in tissues during treatment with solutions of osmium tetroxide are of much more practical use in providing guide lines for the choice of fixatives and fixation methods, although comparative studies have shown considerable variations between one type of specimen and another so that it is difficult to make useful generalisations. In addition, few studies have yet been made of the effects of osmium tetroxide on aldehyde-fixed tissues.

An excellent summary of the results of studies of lipid extraction from tissues is given in the review by Stein and Stein (1971). The total loss of lipid from tissues fixed only with osmium tetroxide can be very high, occurring mainly during dehydration, but this loss can be reduced greatly by the addition of calcium to the fixative (e.g. Strauss and Arabian 1969) and by modifications to the dehydration procedure (see Chapter 4). Although most of the lipid is lost during dehydration, a small amount is extracted by the osmium tetroxide fixative itself and this loss is accompanied by immediate changes in membranes which become freely permeable to small ions and protein molecules (Tormey 1965; Amsterdam and Schramm 1966). At the same time proteins enclosed by membranes remain soluble and are able to diffuse out. Osmium tetroxide fixation destroys the osmotic properties of membranes and consequently the osmolarity of the vehicle of the fixative is not important. Similar results are obtained with fixatives of low and high osmolarity (e.g. Bone and Ryan 1972).

The changes in membranes produced during fixation with osmium tetroxide lead to noticeable morphological effects, such as the shrinkage of nerve myelin (Moretz et al. 1969) and the swelling of isolated cells. This swelling occurs in iso-osmotic fixatives (Bahr et al. 1957; Kushida 1962; Millonig and Marinozzi 1968) and can be prevented by the addition of sodium chloride to make the fixative slightly hypertonic. Sucrose should not be used to adjust the osmolarity of the fixative since sucrose does not inhibit the uptake of water by isolated cells, and sea urchin eggs have been observed to swell even more in a sucrose-containing fixative than in a simple osmium tetroxide solution (Millonig and Marinozzi 1968).

There are many reports in the literature of the beneficial effects of adding calcium ions to osmium tetroxide fixatives for the preservation of a variety of structures, including spindle fibres, nerve myelin, membranes and phospholipids. Many of these improvements appear to be directly related to the decreased extraction of lipids resulting from fixation in the presence of calcium ions. Consequently it is often profitable to add calcium chloride, at a final concentration of 1 to 3 mM, to osmium tetroxide fixatives, although care must be taken not to induce a granular precipitation of proteins (Millonig and Marinozzi 1968) or of other components of the fixative, such as phosphates. Millonig (1973) prefers to add magnesium chloride, rather than calcium chloride, to phosphate-buffered fixatives, because precipitates only then form at high concentrations. To date, however, there is little published work to show whether magnesium has similar beneficial effects on the retention of lipids and the preservation of fine structure.

2.3.2 OSMIUM TETROXIDE SOLUTIONS

Warning: Osmium tetroxide volatilises readily at room temperature. The vapours are very harmful to the eyes, nose and throat. Osmium tetroxide must only be handled in a fume cupboard with good ventilation.

Osmium tetroxide is supplied in the form of yellow crystals in glass ampoules, usually in amounts of 0.1 g or 1.0 g, or as an aqueous solution. Some suppliers use ampoules with pre-scored necks so that a cutting tool is not required. For some fixatives, such as those buffered with phosphates, it is customary to dissolve sufficient solid osmium tetroxide in the buffer to obtain the desired concentration (usually 1%). For other fixatives an aqueous solution (usually 2%) is required and if it is not available commercially it is prepared in the following way:

Method of preparation of osmium tetroxide solutions

To prepare a 2% solution, the label on an ampoule containing 0.1 g or 1.0 g of osmium tetroxide is removed and the ampoule is broken and dropped into a brown glass bottle with a glass stopper. 5 or 50 ml of glass distilled water is then added and the bottle shaken. The bottle must be clean or the solution will quickly become discoloured.

The solution should be prepared at least one day before it is required since osmium tetroxide takes a long time to dissolve, unless a sonicator is available; then the solution can be prepared in a few minutes. It is stable for some months and is best stored on a shelf in a fume cupboard away from direct sunlight. It should not be kept in a refrigerator or other closed space since any leakage of vapour will discolour all the internal surfaces and may well affect the other contents. Osmium tetroxide solutions and vapours can penetrate plastics and so glass containers should be used for storage.

Osmium tetroxide is very expensive and some suppliers (see Appendix) are prepared to pay for used fixative which is then regenerated. Alternatively osmium tetroxide can be reclaimed in the laboratory by the oxidative-distillation procedure described by Jacobs and Liggett (1971). These procedures have the advantage of solving the disposal problems and help to conserve a rare material. If it is necessary to dispose of osmium tetroxide down a sink, very large quantities of running water should be used.

2.3.3 PHOSPHATE BUFFERED OSMIUM TETROXIDE FIXATIVES

Phosphate buffers (§ 2.2.5) are superior to other buffers for primary fixation with osmium tetroxide since they are better able to maintain 'physiological' conditions during the period when the osmium is slowly penetrating into and reacting with the tissues (Millonig 1961). Their only disadvantage is the possibility of granular precipitates occurring when calcium (but not magnesium) is added to the fixative (§ 2.2.4) or when phosphate buffered osmium tetroxide is used as a secondary fixative after fixation with glutaraldehyde (Gil and Weibel 1968; Fig. 2.2). However, such precipitates are rarely observed and phosphate buffers are strongly recommended for general use in both primary and secondary fixatives.

Method of preparation of phosphate buffered osmium tetroxide fixatives

1. Prepare the selected phosphate buffer (§ 2.2.5) at the required pH (usually pH 7.0 to 7.4) and molarity (usually 0.1 M).

2. If necessary, adjust the osmolarity of the buffer with sodium chloride (see § 2.2.5, buffer IV) for primary fixatives, or with sucrose for secondary fixatives.

3. If required, add calcium chloride or magnesium chloride (final concentration of 1 to 3 mM) taking care to avoid the formation of a precipitate.

   (*Note:* Tests for precipitation must be done at the temperature at which the fixative is to be used.)

4. Break an ampoule containing solid osmium tetroxide and drop it into a clean, brown, glass-stoppered bottle. Add sufficient phosphate buffer to give the required final concentration of osmium tetroxide (usually 1 %).

The osmium tetroxide takes some time to dissolve and so the fixative should be prepared at least a day before it is required or a sonicator should be used. The pH of the buffer does not change on the addition of the osmium tetroxide. The fixative is stable for some weeks at room temperature and is best stored in a second air-tight container on a shelf in a fume cupboard away from direct sunlight, and not in a refrigerator.

2.3.4 CACODYLATE BUFFERED OSMIUM TETROXIDE FIXATIVES

It is not customary to use cacodylate buffers (§ 2.2.6) for osmium tetroxide fixatives, but when the primary fixation has been with a cacodylate buffered aldehyde fixative it is convenient to use the same buffer for the subsequent fixation with osmium tetroxide, and cacodylate buffers are being increasingly used for this purpose.

Method of preparation of cacodylate buffered osmium tetroxide fixatives

1. Prepare the cacodylate buffer (§ 2.2.6) at the required molarity (usually 0.1 M) and pH (usually 7.2 to 7.4).

2. If necessary, adjust the osmolarity of the buffer with sucrose.

3. If required, dissolve sufficient anhydrous calcium chloride or magnesium chloride to give a final concentration of calcium or magnesium of 1 to 3 mM.

4. Break an ampoule containing 0.1 g or 1.0 g of solid osmium tetroxide

and drop it into a clean, brown, glass-stoppered bottle. Add sufficient cacodylate buffer to give the required final concentration of osmium tetroxide (usually 1%) and shake the bottle well.

The osmium tetroxide takes some time to dissolve and so the fixative is prepared at least the day before it is required, or a sonicator should be used. The fixative should be prepared and stored in a fume cupboard where it will be stable for several weeks at room temperature.

2.3.5 VERONAL–ACETATE BUFFERED OSMIUM TETROXIDE FIXATIVES

Veronal–acetate buffers are not suitable for primary fixation with osmium tetroxide (see § 2.2.7) but they are satisfactory, although not recommended, for buffering solutions of osmium tetroxide for use as a second fixative where the control of pH is less important.

Palade's veronal-acetate buffered osmium tetroxide fixative

This fixative was developed by Palade (1952) on the basis of comparative studies on various tissues of rats.

Method of preparation of Palade's fixative
(Palade 1952).

1. Prepare a veronal-acetate stock solution with
    sodium veronal (barbitone sodium)   2.89 g
    sodium acetate (anhydrous)   1.15 g
    or   sodium acetate (hydrated)   1.90 g
    distilled water to make   100   ml
    This solution is stable and keeps for some months at 4 °C.

2. Prepare the veronal-acetate buffered fixative with
    2% osmium tetroxide in water   12.5 ml
    (see § 2.3.2.)
    veronal–acetate stock solution   5.0 ml
    0.1 N HCl   5.0 ml (approx.)
    distilled water to make   25.0 ml

The osmium tetroxide solution and veronal-acetate solution are mixed together in a brown, glass-stoppered bottle in a fume cupboard, and then the HCl is added gradually until the required pH (usually 7.2 to 7.4) is reached. The final concentration of osmium tetroxide is 1%. The fixative is stable for some weeks at room temperature and should be stored in a fume cupboard.

Caulfield's veronal-acetate buffered osmium tetroxide fixative

Palade's (1952) fixative is hypotonic to red blood cells and Caulfield (1957) suggested that the osmolarity be adjusted by the addition of sucrose; 4.5 g of sucrose is added to 100 ml of fixative for animal tissues, and 1.5 g for plant tissues.

Zetterqvist's veronal-acetate buffered osmium tetroxide fixative

Rhodin (1954) and Zetterqvist (1956) used Michaelis' veronal–acetate buffer with added salts. Zetterqvist used a Ringer solution and consequently his fixative contains 1 mM calcium chloride, thus making it suitable for secondary fixation with osmium tetroxide for specimens which require a low concentration of calcium to be present.

Method of preparation of Zetterqvist's fixative

(Zetterqvist 1956)

1. Prepare a veronal–acetate stock solution with

sodium veronal (barbitone sodium)	2.94 g
sodium acetate (hydrated)	1.94 g
distilled water to make	100 ml

   This solution is stable and keeps for some months at 4 °C.

2. Prepare a Ringer's solution with

sodium chloride	8.05 g
potassium chloride	0.42 g
calcium chloride	0.18 g
distilled water to make	100 ml

3. Prepare the veronal–acetate buffered fixative with

2% osmium tetroxide in water (see § 2.3.2.)	25.0 ml
veronal–acetate stock solution	10.0 ml
Ringer's solution	3.4 ml
0.1 N HCl	11.0 ml (approx.)

The fixative is prepared in the same way as Palade's fixative (see above) and the pH is adjusted to the required value (usually pH 7.2 to 7.4) with the HCl. The final concentration of osmium tetroxide is 1%. The fixative is stable for some weeks at room temperature and should be stored in a fume cupboard.

Ryter-Kellenberger veronal-acetate buffered osmium tetroxide fixative

Bacterial nuclei are not well preserved by Palade's fixative and after a systematic study Ryter and Kellenberger (1958) suggested the use of a buffer adjusted to pH 6.0 and containing 10 mM calcium chloride. The fixative is prepared by dissolving 0.1 g of solid osmium tetroxide in veronal–acetate buffer III (see § 2.2.7). The 'standard' fixation for bacteria also includes treatment with a solution of uranyl acetate and is described in detail in § 3.9.3.

2.3.6 COLLIDINE BUFFERED OSMIUM TETROXIDE FIXATIVES

Collidine buffered osmium tetroxide fixatives are not suitable as primary fixatives (see § 2.2.8), and *s*-collidine is a most unpleasant substance, but the method of preparation of the fixative is included here for completeness, since it is used by some workers as a secondary fixative.

Method of preparation of collidine buffered osmium tetroxide fixatives

(Bennett and Luft 1959)

1. Prepare the collidine buffer (see § 2.2.8).

2. Prepare the collidine buffered osmium tetroxide fixative with

    2% osmium in water (see § 2.3.2)   2 ml
    collidine buffer   1 ml

Mix the two components in a brown, glass-stoppered bottle. The final concentration of osmium tetroxide is 1.33%; other concentrations are obtained by varying the ratio of the two solutions. The fixative is stable for many days.

2.4   Glutaraldehyde fixatives

Formaldehyde has long been one of the standard fixatives for light microscopy and was used in some early electron microscope studies, while acrolein was investigated as a possible fixative by Luft in 1959. The potentialities of aldehydes as fixatives for electron microscopy were not fully realised, however, until Sabatini et al. (1963) examined their suitability for use in cytochemical studies. The quality of fixation produced by glutaraldehyde, glyoxal, hydroxyadipaldehyde, crotonaldehyde, pyruvic aldehyde, acetaldehyde, acrolein, methacrolein and formaldehyde in cacodylate- or phosphate-buffered solutions was investigated and it was found that none of these were suitable as a general fixative when used alone since the contrast obtained was low and the conventional image of membranes was not obtained. The important observation made by Sabatini et al. (1963) was that if fixation with an aldehyde, particularly glutaraldehyde, was followed by a second fixation with osmium tetroxide the results were as good as the best obtained with osmium tetroxide alone. Subsequently it has been found that this double fixation is excellent for a great variety of specimens with the result that glutaraldehyde, either alone or in combination with paraformaldehyde, is the most widely used primary fixative for electron microscopy at the present time (see Chapter 3). It is particularly useful for preserving the continuity of membrane systems (e.g. Franzini-Armstrong and Porter 1964; Tormey 1964) (Figs. 2.4 and 2.5), for labile cells, such as those in embryonic and pathological tissues, for isolated cells, and for fixation by perfusion (§ 3.3.3). In addition, it provides better preservation of the contents of the endoplasmic reticulum (Fig. 2.1) and of myofibrils (Franzini-Armstrong and Porter 1964) (Fig. 2.6).

Double fixation with glutaraldehyde followed by osmium tetroxide has many advantages over fixation with osmium tetroxide alone. The size of the block of tissue is now much less critical, since relatively large pieces, some millimeters in dimensions, can be immersed in the primary fixative and smaller blocks cut out from the well preserved surface zone for subsequent fixation in osmium tetroxide. In addition, with the exception of certain delicate tissues, specimens can be stored in buffer in the cold for periods up to several months before post-fixation with no obvious alterations in fine structure.

2.4.1 CHEMICAL AND MORPHOLOGICAL EFFECTS OF GLUTARALDEHYDE FIXATIVES

The excellent results obtained when glutaraldehyde is used as a primary fixative are mainly due to the facts that it reacts very rapidly with proteins and that, being a dialdehyde, it stabilises structures by cross-linking before there is any opportunity for extraction by the buffer to occur. In consequence far more of the ground substance of the cytoplasm and of the extracellular matrices is preserved than after primary fixation with osmium tetroxide (Fig. 2.1).

The reactions between glutaraldehyde and components of tissues have not yet been studied systematically; the available evidence is summarised by Millonig and Marinozzi (1968) and Hayat (1970). It is clear that glutaraldehyde is an excellent fixative for proteins, even though it causes some

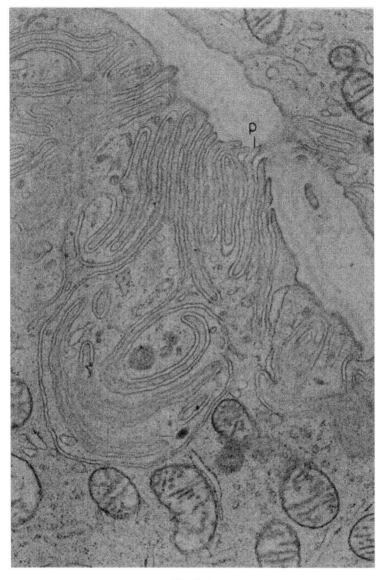

Fig. 2.4a

Fig. 2.4.   Thin sections of rabbit ciliary epithelium. (a) Glutaraldehyde, followed by
osmium tetroxide, fixation. The complex membrane foldings represent interdigitations of
adjacent cells. p, projection from the cell in the right-hand upper corner. (b) Osmium
tetroxide fixation. Many long rows of vesicles are associated with the interdigitations of
adjacent cells. The configuration of the rows is very similar to that of the interdigitations
and arrows indicate places where two rows are joined together to form U-shaped loops in
the manner of interdigitations. These vesicles are interpreted as fixation artefacts.
(From Tormey 1964.)

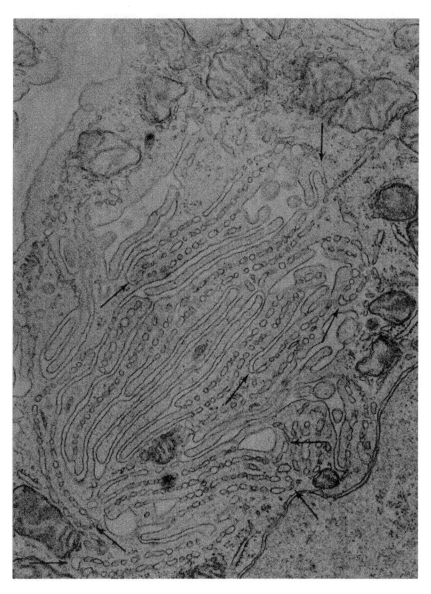

Fig. 2.4b

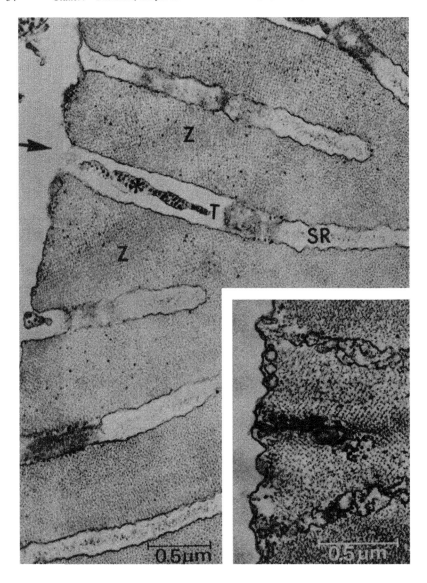

Fig. 2.5.   Transverse sections of fish muscle fibres fixed with glutaraldehyde followed by osmium tetroxide, or (inset) with osmium tetroxide only. The continuities between the transverse tubular (or T) system and the sarcolemma are preserved by glutaraldehyde fixation (arrow). The T system appears in the form of small dense tubules after osmium tetroxide fixation and the continuities with the sarcolemma are disrupted, apparently by the action of the fixative. Z, Z line; SR, sarcoplasmic reticulum. (From Franzini-Armstrong and Porter 1964.)

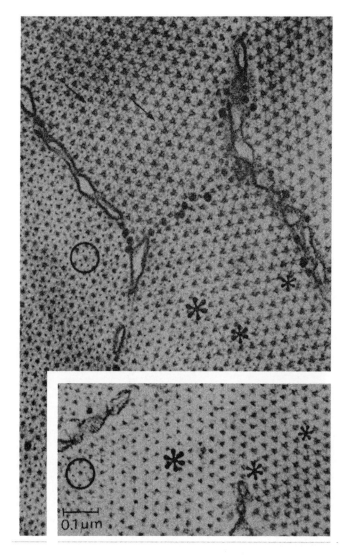

Fig. 2.6. Transverse sections of fish muscle fibres fixed with glutaraldehyde followed by osmium tetroxide, or (inset) with osmium tetroxide only. The circles mark areas of the A bands where the hexagonal arrays of thick and thin filaments are evident. The thick filaments are separated by the same distance in both images but they are obviously larger and more prominent after fixation with glutaraldehyde. Bridges between the filaments are indicated by arrows. The asterisks can be ignored. (From Franzini-Armstrong and Porter 1964.)

significant modifications in their molecular structure and is unable to prevent the much more drastic changes that occur during subsequent treatment with osmium tetroxide (Lenard and Singer 1968). However, it is doubtful whether these molecular changes are necessarily accompanied by morphological effects detectable in electron micrographs of thin sections. From a study of model systems, Millonig and Marinozzi (1968) concluded that the time for the fixation of proteins by glutaraldehyde will be greatly reduced in the presence of monovalent or divalent ions or sucrose. This contrasts with the result of adding sucrose to osmium tetroxide solutions which leads to an inhibition of proper fixation, and suggests that glutaraldehyde is particularly suitable for the fixation of material from sucrose gradients.

Glutaraldehyde reacts with glycogen and after double fixation the density

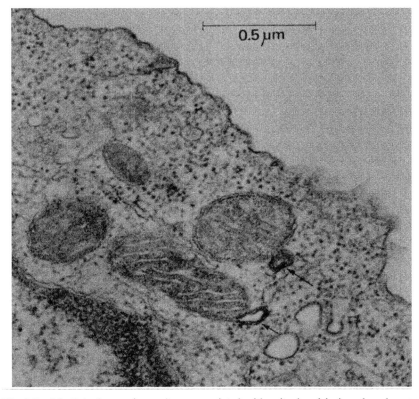

Fig. 2.7.   Myelinic figures (arrows) are associated with mitochondria in a lymphocyte fixed with glutaraldehyde followed by osmium tetroxide. (Unpublished micrograph from a study by Audrey M. Glauert and G. M. W. Cook.)

of glycogen is much higher than after fixation with osmium tetroxide alone, possibly as a result of interactions between osmium and the glutaraldehyde already bound to the glycogen (Millonig and Marinozzi 1968). Lipids are not preserved by glutaraldehyde and unless a second fixation with osmium tetroxide is used as much as 95% of the lipid in a specimen may be extracted during dehydration (Stein and Stein 1971). More seriously, glutaraldehyde and other aldehydes appear to cause phospholipids to pass into solution and myelinic figures (Fig. 2.7) then form when osmium tetroxide interacts with these phospholipids during the second fixation (Curgy 1968). These myelinic figures are seen more frequently in larger blocks of tissue. Fortunately, the formation of myelinic figures can be largely prevented by the addition of 1 to 3 mM calcium chloride to the glutaraldehyde fixative. Calcium also greatly reduces the loss of lipids during dehydration (Mitchell 1969) and improves the preservation of mitochondria (Busson-Mabillot 1971) (Fig. 2.8). As with osmium tetroxide fixatives, care must be taken not to

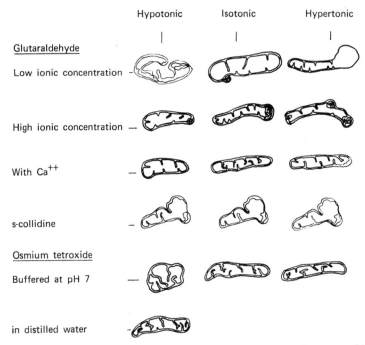

Fig. 2.8. Variations in the fine structure of mitochondria under different conditions of fixation. The addition of low concentrations (0.015–0.05%) of calcium to the fixative reduces swelling and the formation of myelinic figures. The mitochondria appear very swollen after fixation with collidine-buffered glutaraldehyde. (From Busson-Mabillot 1971.)

add too much calcium as this will cause a precipitation of proteins (Millonig and Marinozzi 1968). If problems arise magnesium should be tested as an alternative (§ 2.2.4 and § 2.3.1).

Glutaraldehyde fixation does not destroy the osmotic properties of cells (e.g. Jard et al. 1966), although the cells are frequently changed so that subsequently they are stable only in solutions that are hypertonic or hypotonic relative to the original suspending medium. Glutaraldehyde itself contributes little to the osmotic effects of the fixative and volume changes are not detectable on passing the specimen from the fixative to the same solution without the glutaraldehyde (Bone and Denton 1971). Consequently, the osmolarity of the vehicle determines the effective osmolarity of the fixative. On this basis the majority of the glutaraldehyde fixatives used in electron microscopy are hypotonic, although they are usually described as being hypertonic since the glutaraldehyde itself is assumed to have an osmotic effect (Bone and Ryan 1972). This underlines yet again the impossibility of 'calculating' the 'correct' osmolarity for a fixative or depending on measurements of freezing-point depression.

Glutaraldehyde alone is not an adequate fixative for general use, since certain cell components, especially lipids, are not fixed and may be extracted during dehydration. Consequently, only the results of double fixation with glutaraldehyde followed by osmium tetroxide will be considered here (Fig. 2.9); the appearance of tissues fixed with glutaraldehyde alone was described by Sabatini et al. (1963) in their original paper.

2.4.2 GLUTARALDEHYDE SOLUTIONS

Glutaraldehyde is available commercially, usually as a 25% or 50% solution in water (see Appendix). Most of these solutions contain an impurity which shows spectrophotometric absorption at 235 nm, as compared with 280 nm for monomeric glutaraldehyde, and which has been postulated to be a polymer of glutaraldehyde. It is generally agreed that the monomeric glutaraldehyde is the main reactive species (Gillett and Gull 1972; Korn et al. 1972). In practice commercial solutions of glutaraldehyde, particularly Polysciences 'Biological Grade', which are relatively free of the impurity (Gillett and Gull 1972) have been found to be perfectly adequate for the majority of studies of fine structure. The solutions should be stored at 4 °C or at − 20 °C and barium carbonate should *not* be added. They should be discarded if the pH falls below 3.5 (Sabatini et al. 1964).

For certain special applications, such as cytochemical or immunological studies (see Lewis et al. 1974), it may be necessary to use glutaraldehyde

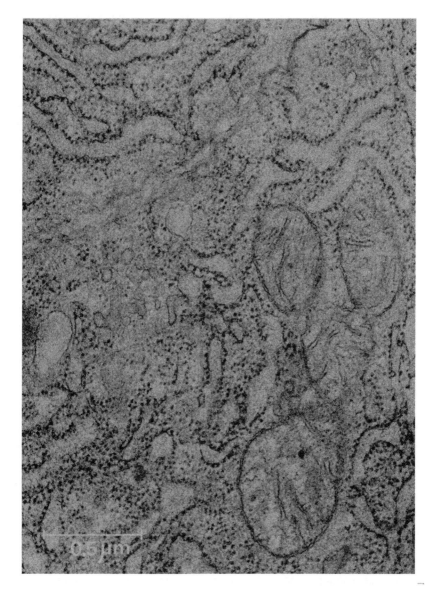

Fig. 2.9. An example of glutaraldehyde, followed by osmium tetroxide, fixation. An electron micrograph of a thin section showing good preservation of rough and smooth membranes, ribosomes and mitochondria in the cytoplasm of an osteoblast. (Unpublished micrograph from a study by Audrey M. Glauert, Honor B. Fell and J. T. Dingle.)

purified by charcoal filtration or distillation. Vacuum distilled glutaraldehyde is available commercially, usually as a 50% solution in water, and is stable for many months if stored at −20 °C. Some commercial samples have a milky precipitate but this is easily removed by filtration (Fahimi and Drochmans 1968).

Vacuum distilled glutaraldehyde is very expensive and so it may be worthwhile purifying it in the laboratory shortly before use. In the simple method described by Smith and Farquhar (1966) the solution is distilled at atmospheric pressure and the distillate collected between 100 and 101 °C in 50 ml aliquots. Samples with a pH of less than 3.4 are discarded and the remainder pooled and bottled. Alternatively, the glutaraldehyde is distilled under vacuum at 15 mm Hg pressure, and the distillate is collected at 80–85 °C and immediately diluted to 25% with boiling water (Gillett and Gull 1972). A milky precipitate may form when the glutaraldehyde is diluted, but this can be prevented by warming the glutaraldehyde at 100 °C for 30 min before dilution (Fahimi and Drochmans 1968).

For purification with charcoal, a sample of commercial 25% glutaraldehyde is shaken with 10% (w/v) activated charcoal at 4 °C for 1 hr and then filtered. The process is repeated 2 or 3 times until a satisfactory UV absorption curve is obtained (Gillett and Gull 1972).

Gillett and Gull (1972) studied the conditions of storage of purified glutaraldehyde and found that temperature was the most important factor. No impurity formed after 8 months at −20 °C and only a small 235 nm peak was observed after 8 months at 4 °C, while large peaks developed after a relatively short time at room temperature and above. Storage in an inert atmosphere (nitrogen) or in the dark had little effect on the rate of deterioration.

The exact concentration of aldehyde groups in commercial samples of glutaraldehyde is not known, but an accurate knowledge is not usually required. If necessary, a measurement of optical density at 280 nm is a simple method of determining the concentration, as well as the osmolarity, of distilled glutaraldehyde (Fahimi and Drochmans 1965a) so long as it is remembered that the absorption is temperature dependent. Alternatively the method devised by Frigerio and Shaw (1969), which is based on the formation of a glutaraldehyde–bisulphite complex followed by iodometric titration of unreacted bisulphite, may be used. This method has a precision of 0.2% or better and is not affected by the presence of buffers so that it can be used directly with fixatives and does not require the distillation of the glutaraldehyde.

2.4.3 PREPARATION OF GLUTARALDEHYDE FIXATIVES

Since cells remain osmotically active during glutaraldehyde fixation the osmolarity of the fixative must be chosen carefully. This is done by adjusting the osmolarity of the fixative vehicle and not of the fixing agent (the glutaraldehyde) which has almost no osmotic effect (§ 2.4.1). The majority of fixatives used in electron microscopy, both for fixation by perfusion and by immersion, are hypotonic (if the effect of the glutaraldehyde is ignored), although they are often stated to be hypertonic since the glutaraldehyde is assumed to contribute to the osmolarity (Bone and Ryan 1972).

Again no general rule can be made in the choice of fixative and the beginner should first follow the procedures used successfully by others with similar specimens and then make modifications if signs of swelling or shrinkage are observed. The osmolarity of glutaraldehyde fixatives is adjusted by altering the osmolarity of the buffer or by the addition of sucrose, glucose or sodium chloride; sucrose does not appear to have any adverse effect on the fixation of proteins by glutaraldehyde (§ 2.4.1) and is generally used.

Although the osmolarity of glutaraldehyde fixatives is important, variations of pH within the range 6.8 to 7.5 rarely appear to have any effect. The pH should be kept below 7.5 to prevent polymerization of the glutaraldehyde and the loss of reactive groups (Sabatini et al. 1964), and for some tissues better fixation is obtained at pH 6.9 to 7.0 than at pH 7.2 to 7.4 (Busson-Mabillot 1971). Veronal-acetate is not a suitable buffer for aldehydes (§ 2.2.7) and collidine gives poor fixation with some tissues (§ 2.2.8). Results with phosphate and cacodylate buffers are very similar, so long as the osmolarity is the same, and a phosphate buffer should be used whenever possible (§2.2.1). It is strongly recommended that a small amount of calcium or magnesium (usually at a final concentration of 1 to 3 mM) be added to glutaraldehyde fixatives to minimise the formation of myelinic figures and other artefacts. Variations in the concentration of glutaraldehyde between 1.0 and 3.5% appear to have little effect (e.g. Fahimi and Drochmans 1965b).

Glutaraldehyde fixatives should be prepared and stored in clean, glass-stoppered bottles. Freshly prepared fixatives should be used whenever possible, although fixatives are stable for some weeks if kept at 4 °C in the dark in a refrigerator. The pH of cacodylate buffered glutaraldehyde may drop slightly during storage and should be tested regularly, while precipitates form in some phosphate buffered fixatives.

Method of preparation of phosphate buffered glutaraldehyde fixative I

1. Prepare a 0.2 M phosphate buffer at the required pH (see § 2.2.5).
2. Prepare the fixative with:

0.2 M phosphate buffer (ml)	50	50	50	50	50
25% glutaraldehyde in H_2O (ml)	4	6	8	10	12
distilled water to make (ml)	100	100	100	100	100
final concentration of glutaraldehyde	1.0%	1.5%	2.0%	2.5%	3.0%

The pH may drop slightly on adding the glutaraldehyde. The final molarity of the buffer in the fixative is 0.1 M; fixatives of higher osmolarity may be prepared by starting with a more concentrated phosphate buffer.

3. If necessary, adjust the osmolarity of the fixative with sucrose, glucose or sodium chloride.
4. If required, add sufficient anhydrous calcium or magnesium chloride to give a final concentration of 1 to 3 mM, taking care to avoid the formation of a precipitate.
   (*Note:* tests for precipitation must be done at the temperature at which the fixative is to be used.)

Method of preparation of phosphate buffered glutaraldehyde fixative II

(Karlsson and Schultz 1965; Maunsbach 1966a)

Prepare the fixative with

$NaH_2PO_4 . H_2O$ (g)	2.98	3.31	1.8	1.8
$Na_2HPO_4 . 7H_2O$ (g)	30.40	33.71	23.25	23.25
NaCl (g)	–	–	5.0	5.0
25% glutaraldehyde in water (ml)	40	40	40	40
Distilled water to make (ml)	1000	1000	1000	2000
Final concentration of glutaraldehyde	2.5%	2.5%	2.5%	1.25%
Osmolarity of buffer (mosmols)	298	320	440	220
pH of the fixative	7.2–7.3	7.4	7.4	7.4

Method of preparation of phosphate buffered glutaraldehyde fixative III

Prepare the fixative with

2.26% $NaH_2PO_4 \cdot H_2O$ in water	64 ml
25% glutaraldehyde in water	8 ml
2.52% NaOH in water, to adjust pH to required value	x ml
distilled water to make	100 ml

The final concentration of glutaraldehyde is 2%.

Method of preparation of cacodylate buffered glutaraldehyde fixatives

1. Prepare a 0.2 M cacodylate buffer at the required pH (see § 2.2.6).
2. Prepare the fixative with

0.2 M cacodylate buffer (ml)	50	50	50	50	50
25% glutaraldehyde in water (ml)	4	6	8	10	12
distilled water to make (ml)	100	100	100	100	100
final concentration of glutaraldehyde	1.0%	1.5%	2.0%	2.5%	3.0%

The pH may drop slightly on adding the glutaraldehyde. The final molarity of the buffer is 0.1 M; fixatives of higher osmolarity may be prepared by starting with a more concentrated cacodylate buffer.

3. If necessary, adjust the osmolarity of the fixative with sucrose, glucose or sodium chloride.
4. If required, add sufficient calcium or magnesium chloride to give a final concentration of 1 to 3 mM calcium chloride in the fixative.

The pH of the buffered fixative may drop slightly during storage and should be checked before use.

Method of preparation of collidine buffered glutaraldehyde fixatives

The fixative is prepared in exactly the same way as cacodylate buffered fixatives, with the substitution of 0.2 M collidine buffer for the 0.2 M cacodylate buffer.

2.5   Formaldehyde fixatives

Formaldehyde was tested as a fixative in the early days of electron micro-scopy but was found to give poorer results than osmium tetroxide and so was little used, except as part of the 'formal-calcium-post-chromation' procedure developed by Baker (1946) for the preservation of lipids (see § 3.9.1). Poor results were also obtained when formaldehyde was used as a primary fixative before osmium tetroxide (Holt and Hicks 1961) due to the presence of considerable amounts of methanol (11 to 16%) in commercial solutions of formaldehyde (Pease 1964). Improved results were obtained with methanol-free formaldehyde prepared from paraformaldehyde (Robertson et al. 1963), which will be referred to as 'paraformaldehyde', although formaldehyde is the actual fixing agent. Even better fixation of some specimens results from the use of a mixture of paraformaldehyde and glutar-aldehyde (Karnovsky 1965). Paraformaldehyde by itself is now only used for certain special applications where its faster rate of penetration and differ-ent reactivity make it preferable to glutaraldehyde. As with other aldehydes the fixation of lipids and membranes is improved by the addition of calcium chloride, and fixation with paraformaldehyde must be followed by a second fixation with osmium tetroxide. Specimens have a similar appearance to those fixed with glutaraldehyde (Trump and Ericsson 1965).

2.5.1   CHEMICAL AND MORPHOLOGICAL EFFECTS OF FORMALDEHYDE

Formaldehyde reacts with proteins but there is less cross-linking than with glutaraldehyde since formaldehyde is mono-functional, being a mono-aldehyde. Also the reaction is slow and reversible (Hayat 1970) so that care has to be taken during washing, and specimens cannot be stored for long periods after formaldehyde fixation before further processing. Formaldehyde reacts with lipids, although the chemistry of the reaction is not understood (Millonig and Marinozzi 1968). It does not react with most polysaccharides, but some glycogen is retained if the fixation is not prolonged.

The main advantage of formaldehyde is that it has a considerably higher rate of penetration than either glutaraldehyde or osmium tetroxide so that large blocks of tissue are well fixed. The permeability properties of mem-branes are partly preserved during formaldehyde fixation (Jard et al. 1966) and the osmolarity of the fixative must therefore be adjusted carefully. There is evidence that formaldehyde itself is not osmotically active (e.g. Bernard and Wynn 1964).

2.5.2 FORMALDEHYDE SOLUTIONS

Commercially available solutions of formaldehyde are unsuitable for electron microscopy because of their content of methanol. Formaldehyde solutions are therefore prepared in the laboratory from powdered paraformaldehyde (trioxy-methylene) just before use.

Method of preparation of methanol-free formaldehyde from paraformaldehyde
Note: It is essential to work in a fume cupboard.

Prepare a 40% solution of paraformaldehyde by dissolving 40 g of paraformaldehyde powder in 100 ml of double distilled water by heating to 65 °C with continuous stirring. Add a few drops of 40% sodium hydroxide and the solution will become clear. The resulting solution of 40% paraformaldehyde is allowed to cool before mixing with the other components of the fixative.

2.5.3 PREPARATION OF PARAFORMALDEHYDE FIXATIVES

Paraformaldehyde fixatives have similar properties to glutaraldehyde fixatives (see § 2.4.3) and so the osmolarity should be adjusted carefully to a value suitable for the particular tissue under study. The choice of buffer is more important than with glutaraldehyde fixatives. Formaldehyde penetrates rapidly but reacts slowly and consequently there is a possibility that specific ions will have an effect before fixation is complete (e.g. Carson et al. 1972). Phosphate buffers should be used if possible, or cacodylate buffers, and a small amount of calcium chloride added. Collidine buffers are not suitable for general use since they cause lysis and extensive membrane destruction (Carson et al. 1972) (Fig. 2.3 and Table 2.1). The fixative should be prepared and stored in a clean, glass-stoppered bottle and is stable for some weeks if kept at 4 °C in the dark in a refrigerator. The pH of the fixative may drop slightly with time and should be tested before use.

Method of preparation of paraformaldehyde fixatives I
1. Prepare a 0.2 M phosphate (or cacodylate) buffer at the required pH (see § 2.2.5 and § 2.2.6).
2. Prepare the fixative with:

0.2 M buffer (ml)	50	50	50	50
40% paraformaldehyde in water (ml)	2.5	5	7.5	10
distilled water to make (ml)	100	100	100	100
final concentration of paraformaldehyde	1.0%	2.0%	3.0%	4.0%

The pH may drop slightly on adding the paraformaldehyde. The final molarity of the buffer is 0.1 M; fixatives of higher molarity are prepared by starting with a more concentrated buffer.

3. If necessary, adjust the osmolarity of the fixative with sucrose, glucose or sodium chloride

4. If required, add sufficient calcium or magnesium chloride to give a final concentration of 1 to 3 mM, taking care to avoid the formation of a precipitate with a phosphate buffer.

Method of preparation of paraformaldehyde fixatives II

Alternatively the fixative is prepared by dissolving the paraformaldehyde directly in the buffer.

1. Add 2.5 g of powdered paraformaldehyde to 50 ml of 0.2 M buffer, and heat to 65 °C with stirring. The resultant solution is cloudy.

2. Add a few drops of 1 N NaOH, with continued stirring, until the solution becomes clear.

3. Allow the solution to cool and then make up to 100 ml with distilled water.

The fixative contains 2.5% paraformaldehyde in 0.1 M buffer. The pH of the buffer will change as the paraformaldehyde dissolves. For example, Millonig and Bosco (1967) found that the pH of a phosphate buffer fell from 7.8 to 7.4 and then remained constant, while the pH of a cacodylate buffer rose from 7.2 to 7.4 and then fell to pH 7.0 after one week. Millonig and Bosco (1967) also found that not all commercial samples of para-formaldehyde dissolve in buffers and that some raise the pH to 10.0 or more.

2.5.4 PARAFORMALDEHYDE FIXATIVES FOR VERY LARGE TISSUE BLOCKS

In exceptional circumstances, such as in the fixation of surgical biopsies, very large blocks of tissue have to be fixed. Even fixatives containing a high concentration of paraformaldehyde do not penetrate sufficiently rapidly unless some agent is added to reduce the density of the tissue. A simple solution to this problem is to use a collidine buffered fixative since the extractive effects of collidine then become an advantage. Winborn and Seelig (1970) tested a collidine buffered 10% paraformaldehyde fixative and obtained adequate penetration of large blocks of tissue approximately 3.5 cm^3 in size.

Method of preparation of collidine buffered paraformaldehyde fixatives
(Winborn and Seelig 1970)

1. Add 10 g of powdered paraformaldehyde to 70 ml of distilled water and heat the solution to 70 °C for 20 min with stirring. The resultant mixture is cloudy.
2. While the solution is still warm, add approximately 6 drops of 1 N NaOH with constant stirring. The solution will become clear.
3. Allow the solution to cool, add then add 2.4 ml of purified *s*-collidine.
4. Add 5 ml of 1 N HCl to adjust the pH to 7.3 to 7.4.
5. Make the solution up to 100 ml with distilled water.

The fixative contains 10% paraformaldehyde in 0.2 M collidine buffer.

2.5.5 PARAFORMALDEHYDE–GLUTARALDEHYDE FIXATIVES

Fixatives containing both paraformaldehyde and glutaraldehyde give better preservation of a wide variety of tissues than either aldehyde alone. In consequence, they are very widely used as primary fixatives at the present time (see Chapter 3). Formaldehyde penetrates tissues much more rapidly than glutaraldehyde and it is thought that the formaldehyde temporarily stabilises structures which are subsequently fixed more permanently by the glutaraldehyde (Karnovsky 1965). In consequence pieces of tissue are well fixed to a much greater depth and myelinic figures are less frequently seen than with glutaraldehyde alone. It is still advisable to add calcium chloride to the fixative whenever possible, and fixation must be followed by a second fixation with osmium tetroxide.

The fixative originally suggested by Karnovsky (1965) consists of 4% paraformaldehyde, 5% glutaraldehyde and 0.05% calcium chloride (approx. 5 mM) in approximately 0.08 M cacodylate buffer, pH 7.2, and is extremely hypertonic (2010 mosmols) if it is assumed that the aldehydes contribute to the osmolarity. Consequently lower concentrations of paraformaldehyde (0.5 to 2.0%) and glutaraldehyde (1.0 to 3.0%) are now used, the lowest concentrations being recommended for fixation by perfusion. However, the evidence that the aldehydes are not themselves osmotically active (§§ 2.4.1 and 2.5.1) raises doubts as to whether a reduction in osmolarity is the cause of the improved fixation. Some workers use distilled glutaraldehyde but there is no evidence that this is essential.

Method of preparation of paraformaldehyde-glutaraldehyde fixatives

1. Prepare a 0.2 M phosphate or a 0.2 M cacodylate buffer (see § 2.2.5 and § 2.2.6) at the required pH. Cacodylate buffer is commonly used at pH 7.4.

2. Prepare 20 ml of a 10% solution of paraformaldehyde by dissolving 2.0 g of paraformaldehyde powder in 20 ml of double distilled water and heating to 60 to 65°C (in a fume cupboard). Add a few drops of 1.0 N sodium hydroxide until the solution becomes clear. Allow the solution to cool before use.

3. Prepare the fixative with

0.2 M buffer	50 ml
10% paraformaldehyde in water	20 ml
25% glutaraldehyde in water	10 ml
distilled water to make	100 ml

    This fixative contains 2% paraformaldehyde and 2.5% glutaraldehyde in 0.1 M buffer. Other concentrations are obtained by using different quantities of the constituents. The pH may change on adding the aldehydes.

4. If necessary, adjust the osmolarity of the fixative with sucrose, glucose or sodium chloride.

5. If required, add sufficient calcium or magnesium chloride to give a final concentration of 1 to 3 mM, taking care to avoid the formation of a precipitate with a phosphate buffer.

2.6   Acrolein and other aldehyde fixatives

2.6.1   ACROLEIN FIXATIVES

Acrolein (acrylic aldehyde) is the most rapidly penetrating of the aldehyde fixatives tested for electron microscopy (Sabatini et al. 1964) and is therefore useful for the fixation of large pieces of tissue and for plant cells and micro-organisms with dense cell walls. It is extremely toxic and must be handled with great care. It solubilises lipids, destroys most enzymatic activities, and causes loss of some microtubules (Schultz and Case 1968), and is only used when formaldehyde is unsuitable. As with other aldehyde fixatives, fixation must be followed by a second fixation with osmium tetroxide.

The acrolein available commercially (see Appendix) is a liquid and is suitable for immediate use. If necessary, the acrolein may be purified by distillation to remove the stabiliser (hydroquinone) and polymerized material

(Carstensen et al. 1971), but there is no evidence that purified acrolein gives better fixation. Acrolein fixatives are usually prepared with 10% acrolein in 0.025 or 0.05 M phosphate buffer. The fixatives are hypertonic and sucrose is not added.

Method of preparation of acrolein fixatives

1. Prepare a 0.2 M phosphate or cacodylate buffer (see § 2.2.5 and § 2.2.6).
2. Prepare the 10% acrolein fixative with

0.2 M buffer (ml)	50	25	12.5
acrolein (ml)	10	10	10
distilled water to make (ml)	100	100	100
final concentration of buffer	0.1 M	0.05 M	0.025 M

3. If required, add sufficient calcium or magnesium chloride to give a final concentration of 1 to 3 mM taking care to avoid the formation of a precipitate with a phosphate buffer.

2.6.2 ACROLEIN–GLUTARALDEHYDE FIXATIVES

Acrolein is also used in combination with glutaraldehyde as an alternative to paraformaldehyde–glutaraldehyde fixatives (§ 2.5.5), and gives similarly excellent results.

Method of preparation of acrolein-glutaraldehyde fixatives

1. Prepare a 0.2 M phosphate or cacodylate buffer (see § 2.2.5 and § 2.2.6) at the required pH.
2. Prepare the fixative with:

0.2 M buffer	50 ml
acrolein	1 ml
25% glutaraldehyde in water	10 ml
distilled water to make	100 ml

This fixative contains 1% acrolein and 2.5% glutaraldehyde in 0.1 M buffer. Other concentrations are obtained by using different quantities of the components. The pH may change on adding the aldehydes.

3. If necessary, adjust the osmolarity of the fixative with sucrose, glucose or sodium chloride.
4. If required, add sufficient calcium or magnesium chloride to give a final concentration of 1 to 3 mM taking care to avoid the formation of a precipitate with a phosphate buffer.

2.6.3   ACROLEIN–PARAFORMALDEHYDE–GLUTARALDEHYDE FIXATIVES

Hayat (1970) reports that fixation with a mixture of acrolein, paraformalde-hyde and glutaraldehyde gives good fixation of a wide variety of animal and plant tissues, and is especially effective for very dense tissues. The fixative contains 10% acrolein (3 ml), 6% paraformaldehyde (5 ml), 10% glutar-aldehyde (6 ml), 0.2 M buffer (5 ml) and distilled water (1 ml). It is prepared by making a fresh solution of 6% paraformaldehyde and then adding the other components.

2.6.4   OTHER ALDEHYDE FIXATIVES

A number of other aldehydes were tested as fixatives for electron micro-scopy by Sabatini et al. (1963) in their original study, including glyoxal, hydroxyadipaldehyde, crotonaldehyde, pyruvic aldehyde, acetaldehyde and methacrolein, and later the properties of malonaldehyde, malialdehyde and succinaldehyde were examined also (Sabatini et al. 1964). None of them give as good preservation of fine structure as glutaraldehyde, but some of them (e.g. hydroxyadipaldehyde) are useful for enzyme cytochemical studies (Lewis et al. 1974) since they are less reactive than glutaraldehyde and have little effect on the activity of many enzymes.

Acetaldehyde must be handled with particular care since it is very volatile and has a pungent vapour. It boils at 20 °C and must be stored and used at 0 °C or below.

2.7   Permanganate fixatives

2.7.1   POTASSIUM PERMANGANATE FIXATIVES

Potassium permanganate was suggested as an alternative fixative to osmium tetroxide by Luft (1956) and was popular for the fixation of botanical specimens and membrane systems in the period before the introduction of aldehyde fixation by Sabatini et al. (1963). Permanganate fixatives were favoured for membranes (e.g. Robertson 1959), since they stand out with great clarity against the surrounding cytoplasm. It was later appreciated that this is a consequence of the extraction of many cell components, including soluble cytoplasmic proteins, ribosomes, neurofilaments, neuro-tubules and myofilaments, during fixation and the subsequent dehydration. These destructive effects are better understood now that more is known about the interactions between potassium permanganate and cell components (Hayat 1970; Riemersma 1970). It has been found that potassium perman-ganate does not stabilise protein gels in model experiments (Bradbury and

Meek 1960), obliterates most of the helical character of isolated proteins in solution (Lenard and Singer 1968) and extracts a large part of the proteins from cells during fixation (Millonig and Marinozzi 1968). Furthermore fixation with potassium permanganate does not prevent the extraction of large amounts of lipid from rat liver (Cope and Williams 1969) or nerve myelin (Moretz et al. 1969) during dehydration, more lipid being extracted than after osmium tetroxide fixation, and comparative studies have shown that some components of membranes visible after osmium tetroxide fixation are no longer detectable after potassium permanganate fixation (Sjöstrand and Elfvin 1962).

RNA is completely extracted from tissues while DNA is preserved (Bradbury and Meek 1960) although potassium permanganate is not a good fixative for chromosomes (Porter and Machado 1960). Glycogen is fixed and stained, but the chemical basis of the reaction has not yet been defined (Millonig and Marinozzi 1968).

The inability of potassium permanganate to preserve many cellular components and the swelling induced in many organelles (Fig. 2.10) makes it unsuitable as a general fixative for electron microscopy. It is still favoured, however, for some botanical studies where its ability to penetrate thick cell walls is an advantage (Van Steveninck 1972). It is of value in the examination of some membrane systems since comparative studies have shown that the continuity of certain membranes is better preserved by potassium permanganate than by other fixatives (Doggenweiler and Heuser 1967; Martin and Rosenberg 1968) although in general the reverse is true. Such observations emphasise the importance of comparing results with different procedures so that systematic artefacts can be detected (see Chapter 1).

Luft (1956) originally suggested the use of a veronal–acetate buffer (§ 2.2.7) for potassium permanganate fixatives, but for many tissues the results are no worse if an unbuffered solution is used (e.g. Mollenhauer 1959). The time of fixation with potassium permanganate often appears to be critical.

Method of preparation of potassium permanganate fixatives
   (Luft 1956)

1.   Prepare a veronal–acetate stock solution with

	sodium veronal (barbitone sodium)	2.89 g
	sodium acetate (anhydrous)	1.15 g
or	sodium acetate (hydrated)	1.90 g
	distilled water to make	100 ml

This solution is stable for some months at 4 °C.

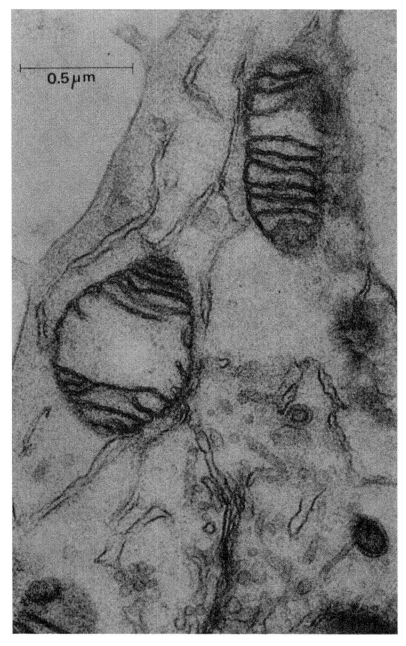

Fig. 2.10.   Section of a rabbit kidney (RK 13) cell fixed with potassium permanganate. Membranes are clearly visible but the mitochondria appear swollen and the ribosomes have not been preserved. (From Dimmock 1970.)

2. Prepare a 1.2% solution of potassium permanganate with

   potassium permanganate ($KMnO_4$)   1.2 g

   distilled water to make   100 ml

   This solution is slowly reduced on contact with air and should be stored in a well-filled, glass-stoppered bottle in a refrigerator.

3. Prepare the potassium permanganate fixative with

   1.2% potassium permanganate   12.5 ml

   veronal-acetate stock solution   5.0 ml

   distilled water   2.5 ml

   0.1 NHCl   5.0 ml

The potassium permanganate solution, veronal–acetate stock solution and distilled water are mixed together in a brown, glass-stoppered bottle, and then the HCl is added gradually until the required pH (usually 7.2 to 7.6) is reached. The final concentration of potassium permanganate is 0.6%; higher concentrations, up to 5%, have been used by some workers and are obtained by starting with a more concentrated solution of potassium permanganate. The fixative is not stable and must be prepared freshly just before use. Millonig and Marinozzi (1968) suggest that 0.5% sodium chloride be added to the fixative to prevent swelling of isolated mammalian cells.

2.7.2 SODIUM PERMANGANATE FIXATIVES

Although potassium permanganate fixation reveals membranes with great clarity, some membranes are disrupted or even destroyed. The observation that plasma membranes of cells in free-living flat worms show discontinuities after potassium permanganate fixation led Wetzel (1961) to test sodium permanganate as an alternative fixative. He stressed the need to maintain a delicate balance between sodium, potassium and calcium ions for the preservation of cell structure and function, and pointed out that disruption of membranes can occur even if the osmolarity of the fixative is correct (see § 2.2.3). The concentration of potassium ions in a 1% potassium permanganate solution greatly exceeds the values generally reported in body fluids, while 1% sodium permanganate does not increase the sodium concentration above that usually found. In a comparison of potassium permanganate and sodium permanganate fixation Wetzel (1961) found that the results obtained were quite similar, although with sodium permanganate there was a slightly, but significantly, better preservation of the integrity and continuity of plasma membranes. Sodium permanganate was equally disruptive to pigment granules and ribosomes.

Method of preparation of sodium permanganate fixative I
(Wetzel 1961)

The 1% sodium permanganate fixative is prepared in exactly the same way as potassium permanganate fixatives (§ 2.7.1). Wetzel (1961) fixed specimens for 2 hr at 1–2 °C in a fixative buffered at pH 6.0.

Method of preparation of sodium permanganate fixative II
(for mammalian tissues)
(Rosenbluth 1963)

The ionic composition of Wetzel's fixative was further modified by Rosenbluth (1963) by the addition of salts to mimic the ionic composition of mammalian and amphibian body fluids.

1. Prepare a solution of 0.83% $NaMnO_4$. $3H_2O$ in veronal–acetate buffer, pH 7.5, by the method used for potassium permanganate (§ 2.7.1).

2. Prepare the sodium permanganate fixative with

NaCl	2.8 g
KCl	0.4 g
$CaCl_2$	0.2 g
$MgCl_2$. $6H_2O$	0.2 g
NaH_2PO_4	0.16 g
0.83% buffered sodium	
permanganate to make	100 ml

The fixative is not stable and is prepared freshly just before use.

Method of preparation of sodium permanganate fixative III
(for amphibian tissues)
(Rosenbluth 1963)

Sodium permanganate fixative II is diluted to 75% of its initial concentration with distilled water. The final concentration of sodium permanganate is about 0.62% and the fixative has the same concentration of sodium, potassium, calcium, magnesium and phosphate ions as an amphibian salt solution.

Rosenbluth (1963) fixed toad spinal ganglia for 1–2 hr at 5 °C and obtained better preservation of slender invaginations of the plasma membrane than with a similarly buffered osmium tetroxide fixative, but a less satisfactory fixation of thin sheets of cytoplasm. Reorganisation of membranes was observed and was thought to be due to physical instability created in the membrane following changes in surface proteins produced by the fixative.

2.7.3 LANTHANUM PERMANGANATE FIXATIVES

Doggenweiler and Frenk (1965) investigated the possibility of using lanthanum permanganate as a fixative for cell membranes on the assumption that La^{3+} would provide an even greater stabilisation of membranes than Ca^{2+}. They found that lanthanum permanganate stains intercellular substances in vertebrate and invertebrate nervous tissue, and a surface layer of plasma membranes (Fig. 2.11). The mechanism of this staining by lanthanum is not yet understood (Lewis et al. 1974), but there is a warning in the study by Dimmock (1970) who showed that the staining is much denser in the presence of anions, such as phosphate, that can form insoluble salts with lanthanum. She recommends that specimens are washed in phosphate-free buffer solution before fixation with lanthanum permanganate.

Lanthanum does not penetrate into cells, and intracellular structures appear the same as after potassium permanganate fixation (Lesseps 1967; Dimmock 1970) (Fig. 2.11).

Doggenweiler and Frenk (1965) prepared lanthanum permanganate from lanthanum sulphate and barium permanganate. A simpler method is described by Lesseps (1967).

Method of preparation of lanthanum permanganate fixatives

(Lesseps 1967)

1. Prepare a veronal-acetate stock solution with

sodium veronal	2.94 g
sodium acetate (hydrated)	1.94 g
distilled water to make	100 ml

   This solution is stable and keeps for some months at 4 °C.

2. Prepare a Ringer's solution (Zetterqvist 1956) with

sodium chloride	8.05 g
potassium chloride	0.42 g
calcium chloride	0.18 g
distilled water to make	100 ml

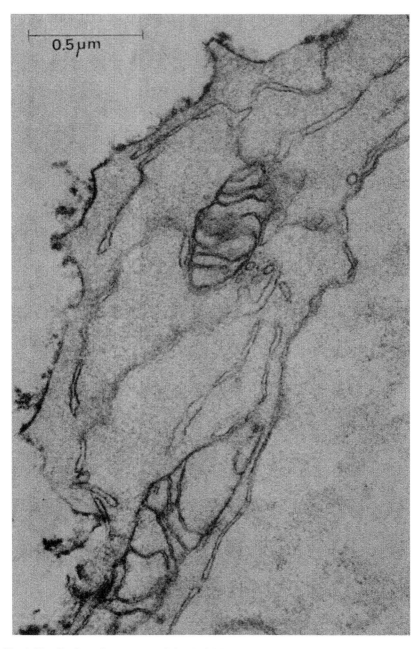

Fig. 2.11.   Section of an RK 13 cell fixed with lanthanum permanganate. The appearance of the cell is similar to that obtained with potassium permanganate fixatives (cf. Fig. 2.10), except for the dense staining of a layer of material at the cell surface.
(From Dimmock 1970.)

3. Prepare the lanthanum permanganate fixative with

$La(NO_3)_3 \cdot 6H_2O$	1 g
$KMnO_4$	1 g
veronal–acetate stock solution	20 ml
Ringer's solution	6 ml
0.1 N HCl	x ml to adjust pH to 7.8
distilled water to make	100 ml

Fragments of tissue or isolated cells are fixed for 1 hr at 0 °C.

2.7.4 OTHER PERMANGANATE FIXATIVES

Comparative studies on the effects of fixation by a range of permanganates have shown that other monovalent salts (lithium and sodium) give results of comparable quality to potassium permanganate (Afzelius 1962, 1963; Hökfelt and Jonsson 1968). Membranes appear very distinct after fixation with calcium, barium or zinc permanganate, and this is thought to be due to the stabilising action of divalent cations on lipids. Cytoplasmic preservation is not good with zinc permanganate fixatives (Afzelius 1962) and some tissues are destroyed (Hökfelt and Jonsson 1968). Doggenweiler and Frenk (1965) reported that barium permanganate produced the same appearance in retina and sciatic nerve as potassium permanganate but with increased contrast.

2.8 Mixed fixatives

Improved preservation of a wider range of structures can often be obtained by mixing two fixing agents together and these mixed fixatives are considered in the present section. Equally good results can often be achieved by using a number of fixatives in sequence as described in the following chapter.

2.8.1 OSMIUM TETROXIDE–POTASSIUM DICHROMATE FIXATIVES

Potassium dichromate neutralised with potassium hydroxide is a strong buffer system in the pH range 5.6 to 7.6 and consequently Dalton (1955) proposed that a mixture of potassium dichromate and osmium tetroxide (chrome–osmium) be used as an alternative to Palade's (1952) veronal–acetate buffered osmium tetroxide fixative (§ 2.3.5). Improved results were obtained with a variety of tissues since there was less leaching out of cytoplasmic materials. This was probably due not only to the presence of a more effective buffer system, but also to the fact that the addition of potassium dichromate accelerates the penetration of osmium tetroxide (Riemersma 1970).

Since the introduction of aldehyde fixatives, chrome–osmium is mainly used as a second fixative as an alternative to osmium tetroxide alone.

Method of preparation of osmium tetroxide-potassium dichromate fixatives
(Dalton 1955)

1. Prepare a buffered solution of 4% potassium dichromate in water with

5% $K_2Cr_2O_7$ in water	80 ml
2.5 N KOH	x ml to adjust pH to 7.2
distilled water to make	100 ml

2. Prepare the fixative with

4% potassium dichromate, pH 7.2	5.0 ml
2% osmium tetroxide (§ 2.3.2)	10.0 ml
3.4% sodium chloride in water	5.0 ml

The fixative contains 1% potassium dichromate, 1% osmium tetroxide and 0.85% sodium chloride, and has a pH of 7.2.

The pH does not change on adding the osmium tetroxide. The fixative is stable for some months at 4 °C.

2.8.2 GLUTARALDEHYDE–OSMIUM TETROXIDE FIXATIVES

Neither glutaraldehyde nor osmium tetroxide is suitable as a general fixative when used alone, and certain artefacts are observed in some specimens when fixation with glutaraldehyde is followed by fixation with osmium tetroxide. These artefacts arise due to the deleterious effects of glutaraldehyde, such as lipid extraction and cell shrinkage, which occur before the osmium tetroxide is applied. Trump and Bulger (1966) showed that most of these effects can be minimised or avoided by simultaneous use of glutaraldehyde and osmium tetroxide in a mixed fixative. The fixative is not stable, since the two fixing agents react with each other. The mechanism of fixation is not known since each of the reagents may be in a different state than when used alone. For example, Hopwood (1970) has shown that the rate of cross-linking of proteins by glutaraldehyde is decreased in the presence of osmium tetroxide, due to competition for amino acid residues.

The fixative suggested by Trump and Bulger (1966) consisted of 6.25% glutaraldehyde and 1.0% osmium tetroxide in collidine buffer. This fixative was later modified by Hirsch and Fedorko (1968) who used a lower concentration of glutaraldehyde and a cacodylate buffer, and followed the

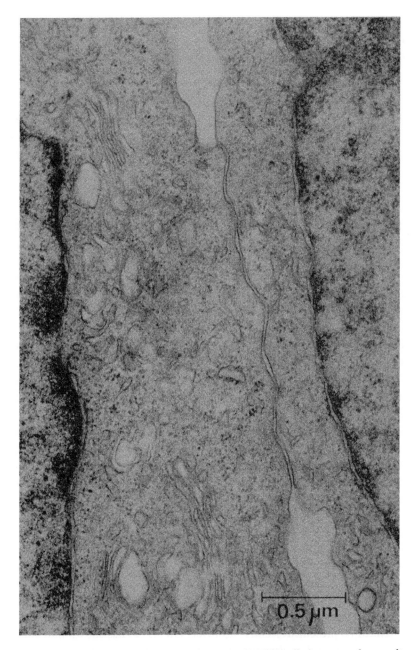

Fig. 2.12. Section of parts of two neighbouring C3 HS/1 cells in a monolayer culture ixed with a mixture of glutaraldehyde and osmium tetroxide. (Unpublished micrograph from a study by Audrey M. Glauert and Mary R. Daniel.)

glutaraldehyde-osmium tetroxide fixation with a second fixation with uranyl acetate (see § 2.9). The osmolarity of the fixative vehicle does not appear to be of importance (Bone and Ryan 1972).

Method of preparation of glutaraldehyde-osmium tetroxide fixatives
(Hirsch and Fedorko 1968)

1. Prepare a stock solution of 2.5% glutaraldehyde in 0.1 M cacodylate buffer, pH 7.4 (§ 2.4.3).
2. Prepare a stock solution of 10% osmium tetroxide in 0.1 M cacodylate buffer, pH 7.4 (§ 2.3.4).
3. Bring the two solutions to 0 °C in an ice bath, and mix one part of glutaraldehyde solution with 2 parts of osmium tetroxide solution.

The fixative consists of about 0.8% glutaraldehyde and 0.7% osmium tetroxide in 0.1 M cacodylate buffer, pH 7.4. The relative concentrations of glutaraldehyde and osmium tetroxide are not critical. Fixation must be carried out at 0 to 4 °C to reduce the rate of reaction between the glutaraldehyde and the osmium tetroxide, and specimens are fixed for a maximum of one hour. At room temperature the fixative turns red-brown after 1 hr and black after 6 hr, and the pH falls (Trump and Bulger 1966).

Cells and tissues fixed with the mixed fixatives have many cytological features intermediate in appearance between those seen with either fixative alone (Fig. 2.12). The mixture is recommended for fragile structures, such as those in pathological cells (Trump and Bulger 1966) and for isolated cells and monolayers of cells (Hirsch and Fedorko 1968).

Presumably mixtures of osmium tetroxide with other aldehydes would give similar results. A mixture of 2% osmium tetroxide and Karnovsky's paraformaldehyde–glutaraldehyde fixative (§ 2.5.5) was tested by Pollard and Ito (1970) and gave comparable preservation of amoebae to aldehyde fixatives containing trinitro compounds (§ 2.8.4).

2.8.3 FIXATIVES CONTAINING POTASSIUM FERRICYANIDE

The addition of potassium ferricyanide to fixatives for the preservation of lipids was first suggested by Elbers et al. (1965) in a study of the fixation of phospholipids by tricomplex 'flocculation'. Subsequently, potassium ferricyanide was added to various fixatives for the preservation of labile lipids, and densely stained material was observed in regions where lipids were thought to be present. Recently, however, Gil (1972) has shown that stained material is still visible after the extraction of phospholipids and concluded that no statements about the chemical nature of the reaction product can be made. In spite of these doubts, the fixatives appear to have some value in morphological studies and so are included here. Various fixation schedules have been used.

Dermer (1969) fixed lung tissue by immersion firstly in 2% glutaraldehyde in cacodylate buffer, pH 7.2, then in a mixture of 0.05 M potassium ferricyanide $(K_3Fe(CN)_6)$ and 0.05 M lead nitrate, and finally in an osmium tetroxide fixative. Finlay-Jones and Papadimitrou (1972) showed that better results were obtained if the initial fixation was done by perfusion.

Bluemink (1972) fixed *Xenopus* eggs by a method developed by De Bruijn (1969). The eggs were immersed firstly in 1% distilled acrolein plus 2.5% glutaraldehyde in 0.067 M cacodylate buffer, pH 7.4 containing 1 mM calcium chloride (for 24 hr at room temperature), then stored overnight in the dark in the same buffer containing calcium chloride, and finally fixed in 1% osmium tetroxide in the same buffer, containing 0.05 M potassium ferricyanide and 0.05 M calcium chloride (for $3\frac{1}{2}$ hr in the dark and cold). The method was successful in preserving labile lipids (Fig. 2.13).

Karnovsky (1971) reported that an osmium tetroxide–potassium ferricyanide fixative greatly increased the contrast of the outer surfaces of membranes and of glycogen.

2.8.4 ALDEHYDE FIXATIVES CONTAINING TRINITRO COMPOUNDS

The effect of adding trinitro compounds to an aldehyde fixative was examined by Stefanini et al. (1967) and Zamboni and De Martino (1967) who used a mixture of paraformaldehyde and trinitrophenol (picric acid) (PAF) for the fixation of spermatozoa and large pieces of tissue. The fixative penetrated rapidly and gave good preservation of whole small embryos and whole organs, such as the kidney. It was very stable, withstanding exposure to light at room temperature for 12 months. Subsequently Ito and Karnovsky (1968) explored the use of a range of trinitro compounds which were added at concentrations of 0.02 to 1.0% to the paraformaldehyde–glutaraldehyde fixative of Karnovsky (1965)(§ 2.5.5) diluted to contain 2% paraformaldehyde and 2.5% glutaraldehyde in 0.1 or 0.2 M phosphate or cacodylate buffer at pH 7.2. Five different trinitro compounds were tested and it was found that 2,4,6-trinitrocresol (FGC), 2,4,6-trinitroresorcinol (FGR) and 2,4,6-trinitrophenol (FGP) give markedly better fixation than 1,3,5-trinitrobenzene (FGB) and 2,4,6-trinitrotoluene (FGT), and that all these fixatives are superior to the paraformaldehyde–picric acid (PAF) fixative used by Stefanini et al. (1967).

Note: All trinitro compounds are potentially *explosive*.

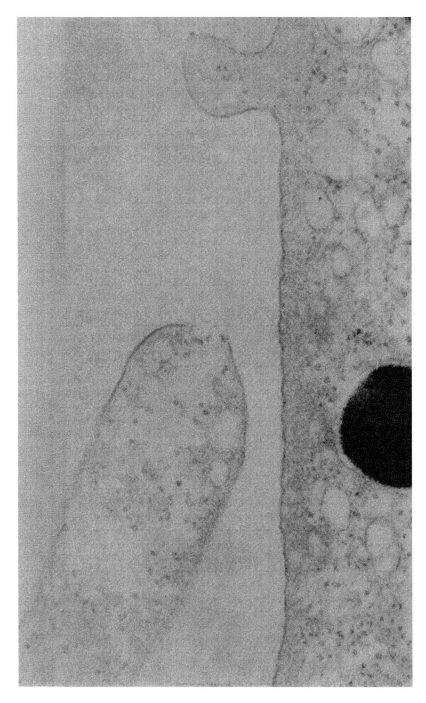

Fig. 2.13a

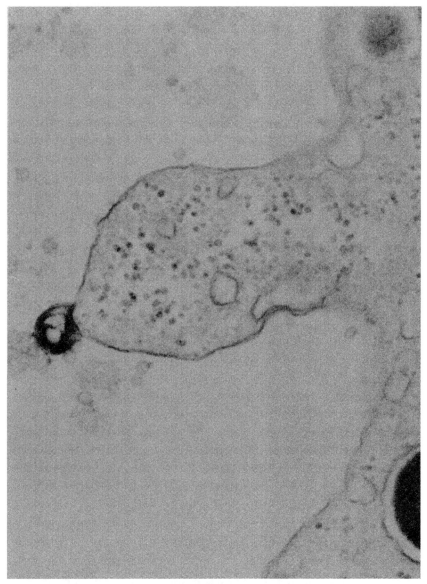

Fig. 2.13b

Fig. 2.13. Sections of surface protrusions in the region of furrow formation in the first cleavage zygote of *Xenopus laevis*. (a) After fixation with glutaraldehyde and acrolein, followed by osmium tetroxide (Bluemink 1970) the membrane at the tip of a surface protrusion shades off into a bleb which is barely visible. (b) After fixation with glutaraldehyde and acrolein, followed by osmium tetroxide and potassium ferricyanide (Bluemink 1972) the bleb at the tip of a surface protrusion consists of material arranged in myelin-like configuration. (Unpublished micrographs from a study by John G. Bluemink.)

After fixation specimens must be washed very thoroughly in buffer (preferably overnight) to remove all the fixative before secondary fixation with osmium tetroxide; small black particles are observed in insufficiently washed tissues.

Ito and Karnovsky (1968) obtained more consistent results with FGC, FGR and FGP, with a variety of tissues fixed by immersion or perfusion, than with paraformaldehyde–glutaraldehyde alone. In particular, membrane systems were well fixed, myelin figures were rare with FGP and FGC, and excellent preservation of the smooth endoplasmic reticulum in steroid-secreting cells was obtained.

Three of these mixed fixatives (FGR, FGC and FGP) were also tested by Pollard and Ito (1970) for the fixation of amoebae and gave satisfactory results, FGC being preferred since it gave the fewest myelin figures. Problems were encountered, however, since the organisms shrank to as little as half their original volume during fixation. If the amoebae were passed rapidly through the fixative to avoid shrinkage, they then shrank during dehydration. In addition, Pollard and Ito (1970) also noted that the amoebae underwent violent abnormal contractions during the first few minutes in the fixative at 22 °C. These contractions did not occur in cold fixative and so it is essential to use the fixative at 0–4 °C.

2.8.5 FIXATIVES CONTAINING DIGITONIN

The addition of 0.2% digitonin to an aldehyde fixative helps to retain free cholesterol which is extracted following convential fixation procedures (Frühling et al. 1969; Scallen and Dietert 1969). This enhanced retention of cholesterol is accompanied by an improvement in the morphological preservation of some types of cell, except for the appearance of many small cylindrical 'spicules' in association with membranes. This artefact can be avoided by using a mixed fixative consisting of 2.5% glutaraldehyde, 1% osmium tetroxide and 0.2% digitonin in a cacodylate buffer at pH 7.4 (Napolitano et al. 1969).

2.8.6 OTHER ADDITIVES FOR FIXATIVES

There are many reports in the literatures of experiments on the effects of adding various compounds to aldehyde fixatives; most of these fixatives are used for special purposes and none of them appear to be suitable for general application. These compounds include potassium permanganate (Kaye and Moses 1960), a nitrogen mustard derivative (Williams and Luft 1968) and potassium dichromate (Robison and Lipton 1969).

Dimethyl sulphoxide has been added to aldehyde fixatives to accelerate penetration (Sandborn et al. 1969; Schwab et al. 1970), but may have a disruptive effect on membranes if the concentration is too high (Winborn and Seelig 1970).

A number of compounds which react with polysaccharides have been added to fixatives, particularly for the fixation and staining of cell surfaces. These include alcian blue (Behnke and Zelander 1970), lanthanum (e.g. Revel and Karnovsky 1967) and ruthenium red (Luft 1971).

Shea and Karnovsky (1969) and Shea (1971) added alcian blue or cetyl-pyridinium chloride to the primary glutaraldehyde fixative, and lanthanum nitrate to the secondary osmium tetroxide fixative and reported an enhancement of the staining obtained by the addition of lanthanum alone to the fixatives. A mixed fixative containing osmium tetroxide and zinc iodide appears to stain certain lipids (e.g. Blümcke et al. 1973), while glutaraldehyde treated with hydrogen peroxide gives improved fixation of a variety of tissues (Peracchia and Mittler 1972) (Fig. 2.14).

The staining properties of fixatives are discussed in detail by Lewis et al. (1974) in another book in this series.

2.9 Uranyl acetate fixatives

In a study of the fixation of bacteria Ryter and Kellenberger (1958) showed that a 'wash' in a solution of uranyl acetate in a veronal–acetate buffer, following osmium tetroxide fixation and preceding dehydration, helped to stabilise the nuclear material. Subsequently, treatment with uranyl acetate became a routine in the fixation of bacteria (§ 3.9.3).

More recently it has been realised that uranyl acetate is also useful as a fixative for animal tissues, being used as a third fixative following fixation with an aldehyde and with osmium tetroxide (e.g. Farquhar and Palade 1965). It is particularly effective in the fixation of membranes (e.g. Brightman and Reese 1969) (Fig. 2.15), as a consequence of the stabilisation of phospholipids (Silva et al. 1968; Hayat 1970; Silva et al. 1971), although there is an accompanying extraction of other components (Farquhar and Palade 1965). For example, glycogen may be extracted or rendered unstainable.

The fixative consists of 0.25 to 2.0% uranyl acetate in water or in a veronal–acetate (Ryter and Kellenberger 1958) or sodium maleate (Karnovsky 1967) buffer. The final pH is usually in the region of pH 5.0. Phosphate and cacodylate buffers cannot be used since precipitation occurs, and tissues fixed in phosphate or cacodylate buffered osmium tetroxide must be

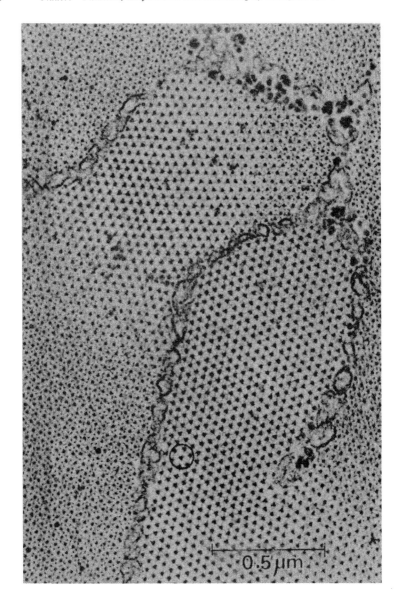

Fig. 2.14.   Transverse section of fish muscle fixed with glutaraldehyde treated with hydrogen peroxide, followed by osmium tetroxide and then uranyl acetate. Compare with Fig. 2.6. (Unpublished micrograph from a study by Clara Franzini-Armstrong.)

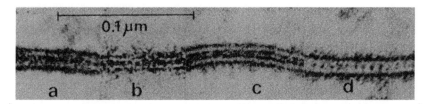

Fig. 2.15. A comparison of the appearance of gap junctions after various preparative procedures. (a) Initial fixation in osmium tetroxide followed by immersion in potassium permanganate after dehydration. (b) Potassium permanganate fixation only. (c) Initial fixation in osmium tetroxide followed by immersion in uranyl acetate before dehydration. (d) Fixation in osmium tetroxide only. Treatment with uranyl acetate before dehydration gives the best preservation of membranes. (From Brightman and Reese 1969.)

washed thoroughly before fixation with uranyl acetate (Farquhar and Palade 1965; Hayat 1970). The fixative should be made up freshly about once a week, and should be stored at room temperature in the dark. The staining properties of the fixative decrease with time and a precipitate gradually forms. The staining effects of uranyl acetate are described by Lewis et al. (1974).

REFERENCES

Afzelius, B. A. (1962), Chemical fixatives for electron microscopy, in: The interpretation of ultrastructure, R. J. C. Harris, ed. (Academic Press, New York and London), p. 1.
Afzelius, B. A. (1963), Experiments with simple fixatives, J. Ultrastruct. Res. *9*, 393.
Amsterdam, A. and M. Schramm (1966), Rapid release of the zymogen granule protein by osmium tetroxide and its retention during fixation by glutaraldehyde, J. Cell Biol. *29*, 199.
Bahr, G. F., G. Bloom and U. Friberg (1957), Volume changes of tissues in physiological fluids during fixation in osmium tetroxide or formaldehyde and during subsequent treatment, Expl Cell. Res. *12*, 342.
Baker, J. R. (1946), The histochemical recognition of lipine, Q. Jl Microsc. Sci. *87*, 441.
Baker, J. R. (1965), The fine structure produced in cells by fixatives, Jl R. microsc. Soc. *84*, 115.
Behnke, O. and T. Zelander (1970), Preservation of intercellular substances by the cationic dye Alcian blue in preparative procedures for electron microscopy, J. Ultrastruct. Res. *31*, 424.
Bennett, H. S. and J. H. Luft (1959), s-collidine as a basis for buffering fixatives, J. biophys. biochem. Cytol. *6*, 113.
Bernard, G. R. and G. G. Wynn (1964), Weight responses of tissue slices and albumin–gelatin gels during formaldehyde fixation with observations upon the effect of pH, Anat. Rec. *150*, 463.
Bluemink, J. G. (1970), The first cleavage of the amphibian egg, J. Ultrastruct. Res. *32*, 142.
Bluemink, J. G. (1972), Cortical wound healing in the amphibian egg: an electron microscopical study, J. Ultrastruct. Res. *41*, 95.
Blümcke, S., W. D Kessler, H. R. Niedorf, N. H. Becker and F. J. Veith (1973), Ultra-

structure of lamellar bodies of type II pneumocytes after osmium–zinc impregnation, J. Ultrastruct. Res. *42*, 417.

Bone, Q. and E. J. Denton (1971), The osmotic effects of electron microscope fixatives, J. Cell Biol. *49*, 571.

Bone, Q. and K. P. Ryan (1972), Osmolarity of osmium tetroxide and glutaraldehyde fixatives, Hist. J. *4*, 331.

Bradbury, S. and G. A. Meek (1960), A study of potassium permanganate 'fixation' for electron microscopy, Q. Jl Microsc. Sci. *101*, 241.

Brightman, M. W. and T. S. Reese (1969), Junctions between intimately apposed cell membranes in the vertebrate brain, J. Cell Biol. *40*, 648.

Busson-Mabillot, S. (1971), Influence de la fixation chimique sur les ultrastructures. 1. Étude sur les organites du follicule ovarien d'un poisson téléostéen, J. Microscopie *12*, 317.

Carson, F., J. A. Lynn and J. H. Martin (1972), Ultrastructural effect of various buffers, osmolality, and temperature on paraformaldehyde fixation of the formed elements of blood and bone marrow, Tex. Rep. Biol. Med. *30*, 125.

Carstensen, E. L., W. G. Aldridge, S. Z. Child, C. P. Sullivan and H. H. Brown (1971), Stability of cells fixed with glutaraldehyde and acrolein, J. Cell Biol. *50*, 529.

Caulfield, J. B. (1957), Effects of varying the vehicle for OsO_4 in tissue fixation, J. biophys. biochem. Cytol. *3*, 827.

Claude, A. (1962), Fixation of nuclear structures by unbuffered solutions of osmium tetroxide in slightly acid distilled water, Proc. 5th Int. Congr. Electron Microscopy, Philadelphia *2*, L-14.

Cope, G. H. and M. A. Williams (1969), Quantitative studies on the preservation of choline and ethanolamine phosphatides during tissue preparation for electron microscopy. II. Other preparative methods, J. Microscopy *90*, 47.

Curgy, J.-J. (1968), Influence du mode de fixation sur la possibilité d'observer des structures myéliniques dans les hepatocytes d'embryons de poulet, J. Microscopie *7*, 63.

Dalton, A. J. (1955), A chrome–osmium fixative for electron microscopy, Anat. Rec. *121*, 281.

Dawson, R. M. C., D. C. Elliott, W. H. Elliott and K. M. Jones (1969), Data for biochemical research, 2nd edition (Clarendon Press, Oxford).

De Bruijn, W. C. (1969), Thesis, Rotterdam.

Dermer, G. B. (1969), The fixation of pulmonary surfactant for electron microscopy. 1. The alveolar surface lining layer, J. Ultrastruct. Res. *27*, 88.

Dimmock, E. (1970), The surface structure of cultured rabbit kidney cells as revealed by electron microscopy, J. Cell Sci. *7*, 719.

Doggenweiler, C. F. and S. Frenk (1965), Staining properties of lanthanum on cell membranes, Proc. Natn. Acad. Sci. U.S.A. *53*, 425.

Doggenweiler, C. F. and J. E. Heuser (1967), Ultrastructure of the prawn nerve sheath. Role of fixative and osmotic pressure in vesiculation of thin cytoplasmic laminae, J. Cell Biol. *34*, 407.

Elbers, P. F., P. H. J. T. Ververgaert and R. Demel (1965), Tricomplex fixation of phospholipids, J. Cell Biol. *24*, 23.

Fahimi, H. D. and P. Drochmans (1965a), Essais de standardisation de la fixation au glutaraldéhyde. I. Purification et détermination de la concentration du glutaraldéhyde, J. Microscopie *4*, 725.

Fahimi, H. D. and P. Drochmans (1965b), Essais de standardisation de la fixation au glutaraldéhyde. II. Influence des concentrations en aldéhyde et de l'osmolalité, J. Microscopie *4*, 737.

Fahimi, H. D. and P. Drochmans (1968), Purification of glutaraldehyde and its significance for preservation of acid phosphates activity, J. Histochem. Cytochem. *16*, 199.

Farquhar, M. G. and G. E. Palade (1965), Cell junctions in amphibian skin, J. Cell Biol. *26*, 263.

Finlay-Jones, J. M. and J. M. Papadimitrou (1972), Demonstration of pulmonary surfactant by tracheal injection of tricomplex salt mixture: electron microscopy, Stain technol. *47*, 59.

Franzini-Armstrong, C. and K. R. Porter (1964), Sarcolemmal invaginations constituting the T system in fish muscle fibers, J. Cell Biol. *22*, 675.

Frigerio, N. A. and M. J. Shaw (1969), A simple method for determination of glutaraldehyde, J. Histochem. Cytochem. *17*, 176.

Frühling, J., W. Penasse, G. Sand and A. Claude (1969), Préservation du cholostérol dans la corticosurrénale du rat au cours de la préparation des tissus pour la microscopie électronique, J. Microscopie *8*, 957.

Gil, J. (1972), Effect of tricomplex fixation on lung tissue, J. Ultrastruct. Res. *40*, 122.

Gil, J. and E. R. Weibel (1968), The role of buffers in lung fixation with glutaraldehyde and osmium tetroxide, J. Ultrastruct. Res. *25*, 331.

Gillett, R. and K. Gull (1972), Glutaraldehyde – its purity and stability, Histochemie *30*, 162.

Hayat, M. A. (1970), Principles and techniques of electron microscopy. Volume 1. Biological applications (Van Nostrand Reinhold, New York).

Hirsch, J. G. and M. A. Fedorko (1968), Ultrastructure of human leukocytes after simultaneous fixation with glutaraldehyde and osmium tetroxide and 'postfixation' in uranyl acetate, J. Cell Biol. *38*, 615.

Hökfelt, T. and G. Jonsson (1968), Studies on reaction and binding of monoamines after fixation and processing for electron microscopy with special reference to fixation with potassium permanganate, Histochemie *16*, 45.

Holt, S. J. and R. M. Hicks (1961), Studies on formalin fixation for electron microscopy and cytochemical staining purposes, J. biophys. biochem. Cytol. *11*, 31.

Hopwood, D. (1970), The reactions between formaldehyde, glutaraldehyde and osmium tetroxide, and their fixation effects on bovine serum albumin and on tissue blocks, Histochemie *24*, 150.

Ito, S. and M. J. Karnovsky (1968), Formaldehyde-glutaraldehyde fixatives containing trinitro compounds, J. Cell Biol. *39*, 168a.

Jacobs, G. F. and S. J. Liggett (1971), An oxidative-distillation procedure for reclaiming osmium tetroxide from used fixative solutions, Stain technol. *46*, 207.

Jard, S., J. Bourguet, N. Carasso and P. Favard (1966), Action de divers fixateurs sur la perméabilité et l'ultrastructure de la vessie de grenouille, J. Microscopie *5*, 31.

Karlsson, U. and R. L. Schultz (1965), Fixation of the central nervous system for electron microscopy by aldehyde perfusion. 1. Preservation with aldehyde perfusates versus direct perfusion with osmium tetroxide with special reference to membranes and the extracellular space, J. Ultrastruct. Res. *12*, 160.

Karnovsky, M. J. (1965), A formaldehyde–glutaraldehyde fixative of high osmolality for use in electron microscopy, J. Cell Biol. *27*, 137A.

Karnovsky, M. J. (1967), The ultrastructural basis of capillary permeability studied with peroxidase as a tracer, J. Cell Biol. *35*, 213.

Karnovsky, M. J. (1971), Use of ferrocyanide-reduced osmium tetroxide in electron microscopy, Proc. 14th Ann. Meeting Am. Soc. Cell Biol. p. 146.

Kaye, G. and M. J. Moses (1960), Combinations of formaldehyde and permanganate as fixatives for light and electron microscopy, Anat. Rec. *136*, 222.

Kelly, A. M. and S. I. Zacks (1969), The histogenesis of rat intercostal muscle, J. Cell Biol. *42*, 135.

Korn, A. H., S. H. Feairheller and E. M. Filachione (1972), Glutaraldehyde: nature of the reagent, J. Molec. Biol. *65*, 525.

Kushida, H. (1962), A study of cellular swelling and shrinkage during fixation, dehydration and embedding in various standard media, J. Electron Microscopy *11*, 135.

Lenard, J. and S. J. Singer (1968), Alteration of the conformation of proteins in red blood cell membranes and in solution by fixatives used in electron microscopy, J. Cell Biol. *37*, 117.

Lesseps, R. J. (1967), The removal by phospholipase C of a layer of lanthanum-staining material external to the cell membrane in embryonic chick cells, J. Cell Biol. *34*, 173.

Lewis, P. R., D. P. Knight and M. A. Williams (1974), Staining methods for thin sections, in: Practical methods in electron microscopy, A. M. Glauert, ed. (North-Holland, Amsterdam).

Luft, J. H. (1956), Permanganate – a new fixative for electron microscopy, J. biophys. biochem. Cytol. *2*, 799.

Luft, J. H. (1959), The use of acrolein as a fixative for light and electron microscopy, Anat. Rec. *133*, 305.

Luft, J. H. (1971), Ruthenium red and violet. 1 Chemistry, purification, methods of use for electron microscopy and mechanism of action, Anat. Rec. *171*, 347.

Malhotra, S. K. (1962), Experiments on fixation for electron microscopy. 1. Unbuffered osmium tetroxide, Q. Jl Microsc. Sci. *103*, 5.

Martin, R. and P. Rosenberg (1968), Fine structural alterations associated with venom action on squid giant axon fibers, J. Cell Biol. *36*, 341.

Maser, M. D., T. E. Powell and C. W. Philpott (1967), Relationships among pH, osmolality, and concentration of fixative solutions, Stain technol. *42*, 175.

Maunsbach, A. B. (1966a), The influence of different fixatives and fixation methods on the ultrastructure of rat kidney proximal tubule cells, I. Comparison of different perfusion fixation methods and of glutaraldehyde, formaldehyde and osmium tetroxide fixatives, J. Ultrastruct. Res. *15*, 242.

Maunsbach, A. B. (1966b), The influence of different fixatives and fixation methods on the ultrastructure of rat kidney proximal tubule cells. II. Effects of varying osmolality, ionic strength, buffer system and fixative concentration of glutaraldehyde solutions, J. Ultrastruct. Res. *15*, 283.

Millonig, G. (1961), Advantages of a phosphate buffer for OsO_4 solutions in fixation, J. appl. Phys. *32*, 1637.

Millonig, G. (1962), Further observations on a phosphate buffer for osmium solutions in fixation, Proc. 5th Int. Congr. Electron Microscopy, Philadelphia *2*, P-8.

Millonig, G. (1964), Study on the factors which influence preservation of fine structure, in: Symposium on electron microscopy, P. Buffa, ed. (Consiglio Nazionale delle Ricerche, Roma), p. 347.

Millonig, G. (1973), Personal communication, Laboratory of Molecular Embryology, Naples.

Millonig, G. and M. Bosco (1967), Some remarks on paraformaldehyde fixation, J. Cell Biol. *35*, 177A.

Millonig, G. and V. Marinozzi (1968), Fixation and embedding in electron microscopy, in: Advances in optical and electron microscopy, Vol. 2, R. Barer and V. E. Cosslett, eds. (Academic Press, New York), p. 251.

Mitchell, C. D. (1969), Preservation of the lipids of the human erythrocyte stroma during fixation and dehydration for electron microscopy, J. Cell Biol. *40*, 869.

Mollenhauer, H. H. (1959), Permanganate fixation of plant cells, J. biophys. biochem. Cytol. *6*, 431.

Moretz, R. C., C. K. Akers and D. F. Parsons (1969), Use of small angle X-ray diffraction to investigate disordering of membranes during preparation for electron microscopy. I. Osmium tetroxide and potassium permanganate, Biochim. biophys. Acta *193*, 1.

Napolitano, L. M., P. R. Sterzing and J. V. Scaletti (1969), Some observations on tissue fixed by glutaraldehyde–osmium tetroxide–digitonin mixtures, J. Cell Biol. *43*, 96a.

Palade, G. E. (1952), A study of fixation for electron microscopy, J. exp. Med. *95*, 285.

Palade, G. E. (1956), The fixation of tissues for electron microscopy, Proc. 3rd Int. Congr. Electron Microscopy, London, p. 129.

Palay, S. L., S. M. McGee-Russell, S. Gordon and M. A. Grillo (1962), Fixation of neural tissues for electron microscopy by perfusion with solutions of osmium tetroxide, J. Cell Biol. *12*, 385.

Pease, D. C. (1964), Histological techniques for electron microscopy, 2nd edition (Academic Press, New York and London).

Peracchia, C. and B. S. Mittler (1972), Fixation by means of glutaraldehyde–hydrogen peroxide reaction products, J. Cell Biol. *53*, 234.

Pollard, T. D. and S. Ito (1970), Cytoplasmic filaments of *Amoeba proteus*, J. Cell Biol. *46*, 267.

Porter, K. R. and R. D. Machado (1960), Studies on the endoplasmic reticulum. IV. Its form and distribution during mitosis in cells of the onion root tip, J. biophys. biochem. Cytol, 7, 167.

Revel, J. P. and M. J. Karnovsky (1967), Hexagonal array of subunits in intercellular junctions of mouse heart and liver, J. Cell Biol. *33*, C7.

Rhodin, J. (1954), Correlation of ultrastructural organization and function in normal and experimentally changed proximal convoluted tubule cells of the mouse kidney, Thesis, Stockholm.

Riemersma, J. C. (1968), Osmium tetroxide fixation of lipids for electron microscopy. A possible reaction mechanism, Biochim. biophys. Acta *152*, 718.

Riemersma, J. C. (1970), Chemical effects of fixation on biological specimens, in: Some biological techniques in electron microscopy, D. F. Parsons, ed. (Academic Press, New York and London), p. 69.

Robertson, J. D. (1959), The ultrastructure of cell membranes and their derivatives, Biochem. Soc. Symp. *16*, 1.

Robertson, J. D., T. S. Bodenheimer and D. E. Stage (1963), The ultrastructure of Maunther cell synapses and nodes in goldfish brain, J. Cell Biol. *19*, 159.

Robison, W. G. and B. H. Lipton (1969), Advantages of dichromate–acrolein fixation for preservation of ultrastructural detail, J. Cell. Biol. *43*, 117a.

Rosenbluth, J. (1963), Contrast between osmium-fixed and permanganate-fixed toad spinal ganglia, J. Cell Biol. *16*, 143.

Ryter, A. and E. Kellenberger (1958), Étude au microscopie électronique de plasmas contenant de l'acide désoxyribonucleique, Z. Naturf. *13*, 597.

Sabatini, D. D., K. Bensch and R. J. Barrnett (1963), Cytochemistry and electron microscopy. The preservation of cellular ultrastructure and enzymatic activity by aldehyde fixation, J. Cell Biol. *17*, 19.

Sabatini, D. D., F. Miller and R. J. Barrnett (1964), Aldehyde fixation for morphological and enzyme histochemical studies with the electron microscope, J. Histochem. Cytochem. *12*, 57.

Sandborn, E. B., T. Makita and K.-N. Lin (1969), The use of dimethyl sulfoxide as an accelerator in the fixation of tissue for ultrastructural and cytochemical studies and in freeze etching of cells. Anat. Rec. *163*, 255.

Scallen, T. J. and S. E. Dietert (1969), The quantitative retention of cholesterol in mouse liver prepared for electron microscopy by fixation in a digitonin-containing aldehyde solution, J. Cell Biol. *40*, 802.

Schultz, R. L. and N. M. Case (1968), Microtubule loss with acrolein and bicarbonate-containing fixatives, J. Cell Biol. *38*, 633.

Schwab, D. W., A. H. Janney and J. Scala (1970), Preservation of fine structures in yeast by fixation in a dimethyl sulfoxide–acrolein–glutaraldehyde solution, Stain technol. *45*, 143.

Shea, S. M. (1972), Lanthanum staining of the surface coat of cells. Its enhancement by the use of fixatives containing alcian blue or cetylpyridinium chloride, J. Cell Biol. *51*, 611.

Shea, S. M. and M. J. Karnovsky (1969), The cell surface and intercellular junctions in liver as revealed by lanthanum staining after fixation with glutaraldehyde with added alcian blue, J. Cell Biol. *43*, 128a.

Silva, M. T., F. C. Guerra and M. M. Magalhães (1968), The fixative action of uranyl acetate in electron microscopy, Experientia *24*, 1074.

Silva, M. T., J. M. S.Mota, J. V. C. Melo and F. C. Guerra (1971), Uranyl salts as fixatives for electron microscopy. Study of the membrane ultrastructure and phospholipid loss in bacilli, Biochim. biophys. Acta *233*, 513.

Sjöstrand, F. S. and L.-G. Elfvin (1962), The layered asymmetric structure of the plasma membrane in the exocrine pancreas cells of the cat, J. Ultrastruct. Res. *7*, 504.

Smith, R. E. and M. G. Farquhar (1966), Lysosome function in the regulation of the secretory process in cells of the anterior pituitary gland, J. Cell Biol. *31*, 319.

Stefanini, M., C. De Martino and L. Zamboni (1967), Fixation of ejaculated spermatozoa for electron microscopy, Nature, Lond. *216*, 173.

Stein, O. and Y. Stein (1971), Light and electron microscopic radioautography of lipids: techniques and biological applications, Adv. Lipid Res. *9*, 1.

Strangeways, T. S. P. and R. G. Canti (1927), The living cell *in vitro* as shown by dark-ground illumination and the changes induced in such cells by fixing reagents, Q. Jl Microsc. Sci. *71*, 1.

Strauss, E. W. and A. A. Arabian (1969), Fixation of long-chain fatty acid in segments of jejunum from golden hamster, J. Cell Biol. *43*, 140a.

Tormey, J. McD. (1964), Differences in membrane configuration between osmium tetroxide – fixed and glutaraldehyde – fixed ciliary epithelium, J. Cell Biol. *23*, 658.

Tormey, J. McD. (1965), Artifactual localization of ferritin in the ciliary epithelium *in vitro*, J. Cell Biol. *25*, 1.

Trump, B. F. and R. E. Bulger (1966), New ultrastructural characteristics of cells fixed in a glutaraldehyde-osmium mixture, Lab. Invest. *15*, 368.

Trump, B. F. and J. L. E. Ericsson (1965), The effect of the fixative solution on the ultrastructure of cells and tissues, Lab. Invest. *14*, 1245.

Van Stevenick, M. E. (1972), Personal communication, University of Queensland.

Wetzel, B. K. (1961), Sodium permanganate fixation for electron microscopy, J. biophys. biochem. Cytol. *9*, 711.

Williams, N. E. and J. H. Luft (1968), Use of a nitrogen mustard derivative in fixation for electron microscopy and observations on ultrastructure of *Tetrahymena*, J. Ultrastruct. Res. *25*, 271.

Winborn, W. B. and L. L. Seelig (1970), Paraformaldehyde and *s*-collidine – a fixative for processing large blocks of tissue for electron microscopy, Tex. Rep. Biol. Med. *28*, 347.

Wood, R. L. and J. H. Luft (1965), The influence of buffer systems on fixation with osmium tetroxide, J. Ultrastruct. Res. *12*, 22.

Zamboni, L. and C. De Martino (1967), Buffered picric acid–formaldehyde: a new, rapid fixative for electron microscopy, J. Cell Biol. *35*, 148A.

Zetterqvist, H. (1956), The ultrastructural organization of the columnar absorbing cells of the mouse jejunum, Thesis, Stockholm.

Chapter 3

Fixation methods

Methods of fixing specimens for electron microscopy are as varied as the specimens themselves and only very general guide lines can be laid down. It is always advisable to first try a method that has been used successfully by other workers with similar specimens and then make modifications as required.

Biological specimens must be fixed as soon as possible after the death of the organism since alterations in fine structure occur rapidly (Fig. 3.1) (Trump et al. 1962). A few special tissues are unusually resistant to post-mortem changes (Ito 1962) but even for these it is advisable to fix the specimens without delay.

Many fixatives, including glutaraldehyde and osmium tetroxide, penetrate tissues slowly and in consequence the central region of a large specimen is rarely well fixed. The vehicle of the fixative usually penetrates more rapidly than the fixing agent itself and may cause damage before the fixation has effectively started. Palade (1956) has described a 'wave of acidification' which precedes the osmium during fixation with unbuffered osmium tetroxide. These damaging effects can be kept to a minimum by making the specimen as small as possible and by careful choice of the vehicle for the fixing agent (§ 2.2.1).

Even when the initial fixation is carried out *in vivo* (§ 3.3.2 and § 3.3.3.) it is usually necessary to complete the fixation by immersion of small pieces of tissue in the fixative. Since the specimens are small and some of the solutions used during fixation are expensive, it is convenient to use small disposable glass vials, about 2 cm in diameter and about 2 cm deep, with close-fitting plastic caps (Fig. 3.2.) for the fixation. In many procedures the specimens can remain in these vials throughout fixation, dehydration and

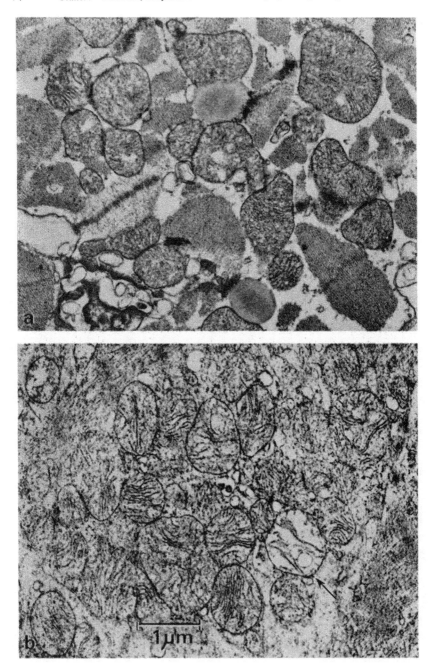

Fig. 3.1.   Transverse sections of rabbit cardiac muscle fixed (a) by perfusion, and (b) immediately after removal from the animal, in a mixture of paraformaldehyde and glutaraldehyde, followed by osmium tetroxide. Post-mortem changes are evident in (b), particularly the swelling of some mitochondria (arrow).

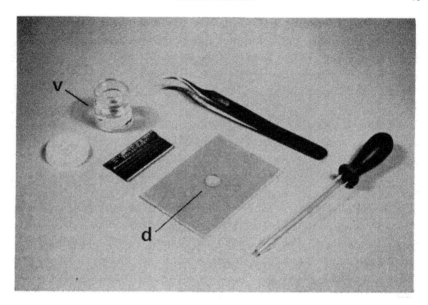

Fig. 3.2. The equipment required for fixation by immersion. Pieces of tissue are placed in a small drop of fixative (d) on a sheet of dental wax, and are then cut into small fragments with a sharp razor blade. These fragments are transferred to a small glass vial (v) containing fixative using a wide-mouthed pipette or pair of tweezers.

impregnation with the embedding medium, and need only be removed when placed in the capsules or moulds for the final embedding.

3.1 Time and temperature of fixation

3.1.1 PRIMARY FIXATION

The optimum time and temperature for primary fixation depends on the nature and size of the specimen and on the composition of the fixative. For isolated cells a few minutes fixation may be sufficient, while for compact tissues several hours may be required, particularly at low temperatures.

In general the time and temperature of fixation should be chosen so that the stabilising action of the fixative is complete, but fixation should not be continued beyond that point or excessive extraction by the buffer may occur. Fixation for 2 hr at 0 °C is common for primary fixation with osmium tetroxide, and similar times and temperatures are frequently used for primary aldehyde fixatives, in spite of the fact that they act more quickly, either as a result of their faster rate of reaction (e.g. glutaraldehyde), or faster rate of penetration (e.g. formaldehyde) or both (e.g. acrolein). Consequently, much shorter times could often be used with safety, and certainly the fixation should not be prolonged since aldehydes do not stabilise lipid components and these may well be extracted during an extended fixation.

Variation of the temperature of fixation with aldehydes or osmium tetroxide between $0\,^{\circ}C$ and $25\,^{\circ}C$ has little observable effect on fine structure, except for certain cell components, such as microtubules, which may not be preserved at low temperatures (e.g. see Gardner et al. 1969). At higher temperatures shrinkage of mitochondria and granularity of the cytoplasm have been reported (Rhodin 1954). Volume changes and extraction are reduced by lowering the temperature to $0\text{–}4\,^{\circ}C$ and this also has the advantage of slowing down enzymatic reactions and thus decreasing the extraction due to autolysis, although the rate of fixation is reduced at the same time. A short preliminary fixation in the cold, followed by the main fixation at room temperature has been recommended for osmium tetroxide (Pease 1964), glutaraldehyde and potassium permanganate (Millonig and Marinozzi 1968) fixation, to minimise extraction and to reduce the overall time of fixation.

3.1.2 WASH

When more than one fixation is being used the specimens are usually washed after the primary fixation to remove all unreacted fixative. It has been suggested that adequate washing is particularly important when aldehyde fixation is followed by osmium tetroxide fixation because the two fixatives react with each other (§ 2.8.2), but there is little published evidence of the adverse effects of a short wash, or none at all. For a few special specimens it is necessary to avoid washing; for example, Bodian (1970) found that storage for as little as 30 min in cacodylate buffer containing sucrose had a flattening effect on certain synaptic vesicles. Artefacts introduced during washing can be avoided by using a combined glutaraldehyde–osmium tetroxide fixative (§ 2.8.2).

It is usual, however, to use at least three changes of washing solution over a period of 2 hr. After glutaraldehyde fixation it is often convenient to store specimens in the washing solution overnight before further processing; this cannot be done after formaldehyde fixation because the reaction of formaldehyde with tissue components is reversible and extraction occurs (§ 2.5.1). For the majority of specimens the vehicle used in the primary fixative is a suitable washing solution, but sucrose may have to be added to adjust the osmolarity for osmotically-sensitive specimens. The washing solution may have to have a different osmolarity to the fixative because the osmotic properties of membranes often change during fixation (§ 2.2.3). Specimens are washed at the same temperature as the primary fixation unless they are to be stored for long periods when it is advisable to keep them in the cold.

3.1.3 SECONDARY FIXATION

The time and temperature for the second fixation are much less critical than for the first. Most commonly specimens are fixed in 1% osmium tetroxide in the same buffer used for the primary fixation for 1 hr at 4 °C or at room temperature.

3.2 Fixation schedules

The sequence of fixation with an aldehyde, followed by osmium tetroxide and possibly also uranyl acetate, has become the standard fixation schedule at the present time for a great variety of specimens.

A representative fixation schedule is:

1. *First fixation* for 1 hr at 4 °C in 2% paraformaldehyde plus 2.5% glutaraldehyde in 0.1 M phosphate buffer, pH 7.4, containing 2.5 mM calcium chloride.
2. *Wash* for 2 hr (or overnight) at 4 °C in 3 changes of 0.1 M phosphate buffer, pH 7.4, containing 2.5 mM calcium chloride.
3. *Second fixation* for 1 hr at room temperature in 1% osmium tetroxide in 0.1 M phosphate buffer, pH 7.4, containing 2.5 mM calcium chloride.

Alternatively, when specimens are to be fixed or 'block stained' with uranyl acetate a different schedule is used because uranyl ions form precipitates in phosphate and cacodylate buffers.

1. *First fixation* for 1 hr at 4 °C in 2% paraformaldehyde plus 2.5% glutaraldehyde in 0.1 M phosphate buffer, pH 7.4, containing 2.5 mM calcium chloride.
2. *Wash* for 2 hr (or overnight) at 4 °C in 3 changes of 0.1 M phosphate buffer, pH 7.4, containing 2.5 mM calcium chloride.
3. *Second fixation* for 1 hr at room temperature in 1% osmium tetroxide in veronal–acetate buffer, pH 7.4, containing 2.5 mM calcium chloride.
4. *Rinse* in veronal–acetate buffer, pH 7.4.
5. *Third fixation* for 1 hr at room temperature in 0.5% uranyl acetate in veronal–acetate buffer, pH 5.0.

3.3   Organs and tissues

3.3.1   IMMERSION FIXATION

Organs and tissues, such as the skin, that can tolerate a temporary inter-
ruption to their blood supply and still retain their function and structure,
can be fixed by immersion *in vitro*. The animal is killed and then the tissue
is removed by dissection as rapidly as possible. A relatively large piece of
tissue is taken and then cut into smaller pieces for fixation, the size of the
piece depending on the nature of the fixative and the density of the tissue.
For primary fixation with osmium tetroxide the tissue is cut into small
cubes which must not be larger than 0.5 mm on each side if the whole cube
is to be well fixed, while for aldehyde fixatives the pieces of tissue can be
considerably larger. Small cubes are then cut later from selected regions of
these pieces while they are in the buffer wash before osmium tetroxide
fixation. The proportion of the larger pieces that is well fixed depends upon
the aldehyde fixative; for glutaraldehyde, which penetrates slowly, the zone
of adequate fixation may well be only 0.5 mm deep, while for a fixative
containing formaldehyde this zone will be considerably deeper, but will
probably not be more than a few millimetres (Hayat 1970). In general, the
small cubes should be cut from as near the surface of the piece of tissue as
possible, while avoiding the surface itself which will have been damaged
when the piece was cut out. The fact that the cubes can be selected in this
way is a great advantage of the double fixation with an aldehyde followed
by osmium tetroxide, as compared with osmium tetroxide alone. A relatively
large proportion of a small cube cut from the fresh tissue for osmium
tetroxide fixation is damaged during cutting, while after aldehyde fixation
the tissue is much tougher and the mechanical damage produced during
cutting into cubes is considerably less.

   A survey of the recent literature indicates that the most popular primary
fixatives for immersion fixation of a wide range of specimens are either
1.7 to 4% glutaraldehyde in 0.1 M phosphate or cacodylate buffer, pH
7.2–7.4 (§ 2.4.3), or diluted Karnovsky's (1965) fixative, consisting of 2%
paraformaldehyde, 2.5% glutaraldehyde and 2.5 mM calcium chloride in a
similar buffer (§ 2.5.5).

   The cutting of the tissues into pieces or cubes is conveniently done by
placing the tissue on a sheet of dental wax in a fume cupboard, adding
enough fixative to keep the tissue moist, and then cutting the tissue cleanly
with a new, single-edged razor blade or dissecting knives (Fig. 3.2). Great
care is required during cutting to ensure that the tissue is damaged as little

as possible. As soon as the pieces or cubes are prepared they are transferred to pre-labelled glass vials containing the fixative (Fig. 3.2). The partially-fixed tissue is easily damaged and is best transferred using a wide-mouthed Pasteur pipette or pieces of filter paper.

The relative volumes of the specimen and the fixative are important; if there is too little fixative the effective concentration of the fixative is reduced by dilution with soluble components of the specimen. In general, the fixative should be at least 10 times greater in volume than the specimen. If fixation is carried out at 0–4 °C the vials containing the fixative are placed in a bath of crushed ice some time before fixation is due to start. Specimens are fixed for 30 min to 1 hr at room temperature or for 1 to 4 hr at 0–4 °C.

After primary fixation the fixative is removed with a fine pipette and flushed down the sink in the fume cupboard with a large quantity of water. The fixative is replaced with the first washing solution (§ 3.2) which is usually the same buffer as used for the fixative with the addition of sucrose, if necessary. The pieces of tissue are then removed from the vials, cut into small cubes not greater than 0.5 mm on each side, and replaced in the vials with fresh washing solution. At this stage the fixed tissue is usually tough enough to be handled with a fine pair of forceps, although great care should be taken at all stages to be as gentle as possible. The chosen fixation schedule is followed (§ 3.2), the specimens remaining in the glass vials throughout the procedure. When fixation or washing is being done in the cold, the vials are transferred to a refrigerator, but all changes of fixative must be done in a fume cupboard.

Marks and Briarty (1970) describe the use of a disposable syringe fitted with a filter as an alternative to a small vial for processing specimens through fixatives and dehydrating agents.

3.3.2 *IN VIVO* FIXATION

The fine structure of many tissues is critically dependent on a continuous blood supply and it is essential that fixation of such tissues should be started while the animal is still alive. For fixation *in vivo* the animal is anaesthetised, the tissue exposed by dissection, and then the fixative is applied by dripping it over the surface of the organ or by injecting it into the tissue.

The dripping method is only useful when the surface layers of an organ are being studied since the fixative does not penetrate through the surface capsule and any attempt to remove the capsule before fixation starts may damage the structure of the organ (Maunsbach et al. 1962). For fixation by this method the cold fixative is dripped onto the surface of the organ continuously for 10 to 20 min, the excess fixative being collected with a suitably placed pad of cotton wool. The method has mainly been used with veronal–acetate buffered osmium tetroxide fixatives (§ 2.3.5) at concentrations of 1–2%. Since it is limited to the study of surface layers of organs it has largely been supplanted by techniques in which the fixative is injected into the tissue, or by fixation by perfusion (§ 3.3.3).

For the examination of the internal structure of organs the fixative is injected into the tissue using a micropipette or hypodermic needle. The fixative is injected slowly for 2 to 15 min and is usually cooled to 0–4 °C. Some typical references are listed in Table 3.1.

After initial fixation *in vivo*, by dripping or injection, the organ is removed and cut into small pieces and the primary fixation completed by immersion in the fixative in the cold or at room temperature (see § 3.3.1), the total fixation time varying from 1 to 4 hr.

These methods of fixation, in common with perfusion fixation, have the disadvantage that the anaesthetic may have an effect on the fine structure of the tissue under study; these effects are reduced to a minimum by starting the fixation as soon as possible after the administration of the anaesthetic. The possible adverse effects of anaesthetics can be avoided by killing the animal by a blow on the head, but it is then necessary to work extremely rapidly since the heart will not continue to beat for long.

3.3.3   PERFUSION FIXATION

The method of *in vivo* fixation described in the previous section (§ 3.3.2) is usually adequate for small organs, or when the surface regions of larger organs are being studied, but poor fixation is obtained of internal regions of organs such as the kidney and the lung, which have a copious blood supply, and of the brain and other tissues of the central nervous system. For these tissues it is necessary to perfuse the fixative through the living, anaesthetised animal.

There are many methods of perfusion recorded in the literature, and probably many more unpublished variants peculiar to a particular laboratory. For convenience of description perfusion procedures will be discussed under three headings: the type of anaesthesia, the route by which the fluids are injected, and the composition of the perfusion fluids.

3.3.3.a.   *Anaesthesia*

The type of anaesthetic used is usually determined by the preferences of the investigator and is likely to be important only in specific physiological

TABLE 3.1

In vivo fixation by injection

Organ	Site of injection	Fixing agent	Vehicle	pH	Temp.	Time	Reference
Bullfrog heart	Sinus	0.88% G	0.067 M cac	7.2	0°C	?	Baldwin (1970)
Mouse diaphragm	Pleural and abdominal cavities	2% P + 2% G	0.09 M cac + 0.025% CaCl₂	7.4	r.t.	3–5 hr	Karnovsky (1967)
Rat pituitary	Cranial cavity	(a) 1% OsO₄	phosphate	7.6	?	few min	Smith and Farquhar (1966)
Rat pituitary	Cranial cavity	(b) 1.5% G	0.067 M cac + 1% sucrose	7.4	?	few min	
Rabbit cartilage	Knee joint	3% G	phosphate + CaCl₂	7.3	?	few min	Palfrey and Davies (1966)
Rat kidney	Lumen of tubule	3% G	0.1 M cac	7.2	0°C	2 min	Maunsbach (1966)
Goldfish brain	Ventricle of heart	1.5% KMnO₄	1.5% NaHCO₃ + salts	7.4	r.t.	approx. 10 min	Robertson (1963)
Bat stomach	Stomach	(a) 1% OsO₄	VeAc	7.1–7.5	cold	few min	Ito and Winchester (1963)
Bat stomach	Stomach	(b) 1.33% OsO₄	collidine + 0.25 M sucrose	7.1–7.5	cold	few min	
Rabbit eye	Posterior chamber	1% OsO₄	VeAc + salts	7.2–7.6	cool	15 min	Tormey (1963)
Cat pancreas	Pancreas	(a) 1% OsO₄	VeAc	7.2–7.4	2–5°C	10–15 min	Sjöstrand and Elfvin (1962)
Cat pancreas	Pancreas	(b) 0.6% KMnO₄	VeAc	7.2–7.4	2–5°C	10–15 min	

Key: G, glutaraldehyde; P, paraformaldehyde; VeAc, veronal-acetate; r.t., room temperature; cac, cacodylate.

or pharmacological studies. Ether and halothane are the most commonly used of the volatile anaesthetics and are particularly convenient for the smaller laboratory animals, such as rats, mice and guinea pigs. For larger animals it is usually more convenient to give an intra-peritoneal injection, which is easy, or an intravenous injection, which is more difficult, and to use nembutal, chloral hydrate or urethane. Until the operator has gained some skill at perfusion a very deep level of anaesthesia should be aimed at; administration of an overdose should do no harm. The dose of anaesthetic to use can be found by consulting a standard laboratory manual (such as Lumb 1963).

3.3.3b.   *The route of injection of perfusion fluids*
The route by which the perfusion fluid is injected is determined largely by the size of the animal, the tissue to be perfused and the surgical skill of the operator. The commonest method of controlling the rate of perfusion is to use a simple gravity feed, but for small laboratory animals it is often sufficient to use a manually operated injection syringe which can be safely refilled several times during the perfusion, provided that care is taken not to dislodge the perfusion needle or to introduce air bubbles. For more sophisticated experiments a perfusion pump can be used (e.g. Gil and Weibel 1969–70).

The perfusion pressure must be high enough to ensure good flow of the fixative but not significantly higher than the arterial pressure in the intact animal. With an intra-arterial gravity feed, the level of the reservoir of perfusion fluid should be 120–150 cm (4–5 ft) above the animal, but when an intravenous route is used, as in perfusion of the liver via the portal vein, the fluid level should be only 20–30 cm above the animal. Animals should always be perfused in a good fume cupboard or under an extraction hood, and the operator must wear protective goggles.

For small animals the most satisfactory route for perfusion is through the left ventricle into the base of the aorta (see below). For young animals a more practicable method is simply to insert a fine hypodermic needle connected to a reservoir of perfusion fluid through the wall of the left ventricle and hold it in place while fixation occurs. As soon as the needle is in place the right atrium should be cut with a fine pair of scissors to allow venous blood and perfusate to escape.

For perfusion of mature rats the following procedure is recommended:
1. When the animal is deeply anaesthetised, lay it on its back in a shallow tray.

2. Make a midline incision down the whole length of the thorax and much of the abdomen and reflect the skin.
3. Expose the heart by cutting the ribs and intercostal muscles along each side and raising the flap so formed.
4. Clean away the thymus and excess adipose tissue from the arch of the aorta and draw a ligature beneath it ready to tie the perfusion cannula in place.
5. Make a stitch through the apex of the heart and clip the thread to the side of the animal so that the heart is under slight tension.
6. Make a small perforation in the wall of the fourth ventricle with fine scissors and quickly insert a cannula into the opening and through the aortic valve so that the cannula projects about 5 mm into the aorta.
7. Tie the cannula in place and as soon as the perfusion fluid is flowing cut the right atrium.

For mice it may prove more convenient to use a cannula made out of plastic tubing. Force a short length of tubing just small enough to enter the aorta into wider tubing so that about 3 mm is projecting. When the tubing is passed through the ventricle, the wider tubing will not enter the aortic valve and the fine tubing will project just far enough to provide an adequate seal.

The method of perfusion described above is very wasteful of fixative in large animals and, if tissues are only required from the head end, the abdominal aorta should be clamped off as high up as possible once the perfusion has begun. For some purposes, in fact, it may be more satisfactory to perfuse via the abdominal aorta in animals larger than the rat, and it is essential to use this route in order to perfuse the heart properly. When perfusing through the aorta it is advisable to clip the aorta in two places and then insert the needle between. To release blood it is best to cut the jugular vein and then the inferior cava.

3.3.3c *Fluids and fixatives for perfusion*
There is no general agreement in the literature about the optimum temperature of the perfusion fluid, or whether the blood should be washed out with saline before the fixative is introduced. For preliminary experiments it is sufficient to use fixative alone at room temperature. If saline is used, a convenient way of ensuring that some saline precedes the fixative in perfusing a small animal is to fill the tubing with saline before carefully filling the reservoir with fixative. For larger animals it is necessary to have

two reservoirs, one with saline and the other with fixative, connected to the perfusion tubing via a three-way tap.

For small animals, where the perfusion rate is low, the temperature of the perfusion fluid can be adequately controlled by passing it through a coil immersed in a water bath. For larger animals the fluid in the reservoir must be kept at an appropriate temperature. This is done most easily by insulating the reservoir with a jacket of expanded polystyrene or similar material.

Fixation is undoubtedly more rapid and effective the higher the temperature. For purely morphological studies, therefore, the perfusion fluid should probably be near the deep body temperature of the animal. For enzyme histochemistry the perfusion fluid should be at room temperature, and for very labile enzymes it may be necessary to cool the fixative down to near 0 °C (Lewis et al. 1974). Perfusion fluids below body temperature are liable to cause vasoconstriction which can impair the effectiveness of the perfusion.

Opinions differ on the optimum concentration of the fixing agent in the initial perfusing fluid. Peters et al. (1968) and Brightman and Reese (1969) followed the lead of Reese and Karnovsky (1967) in using a small volume of a dilute aldehyde fixative first. In contrast, Schultz and Case (1970) specifically blame swollen mitochondria on slow initial fixation and they advocate the use of a small volume of very concentrated fixative (e.g. 19% glutaraldehyde) to precede the main perfusion fluid. With a concentrated initial fixative, a small volume of a saline solution must be used first to wash out the blood. Again a compromise may be necessary; when tissue is to be used for enzyme histochemistry a dilute fixative may have to be used, while for purely morphological studies rapid initial fixation is probably best.

Bohmann and Maunsbach (1970) studied the effect of adding dextran or polyvinylpyrrolidone (PVP) to the fixative to raise the colloid osmotic pressure and reported much less expansion of extravascular spaces. The dextran and PVP which they used had molecular weights of about 40,000 daltons and were at a final concentration of 2–4% in the perfusion fluid. Schultz and Case (1970) used 0.5% or 1% gum acacia in their perfusion fluids; they were troubled by turbidity when gum acacia was present and cured it by the addition of a small quantity of polyphosphate. It seems likely, however, that the turbidity was caused by the very high concentration of phosphates in their perfusion fluid.

For purely morphological studies the following procedure should be suitable for preliminary experiments. It is a simplified modification of the method described by Schultz and Case (1970) and the volumes quoted are for perfusion of an adult rat through a cannula inserted via the left ventricle into the root of the aorta.

1. Begin the perfusion with 10 ml of an isotonic buffered saline solution containing 2% PVP or dextran (of molecular weight in the range 40,000 to 50,000 daltons) or 1% gum accacia. For critical experiments a complete Tyrode solution should be used, but for many purposes some of the minor constituents can be omitted.

2. Continue the perfusion with 15–18 ml of a concentrated glutaraldehyde solution containing:

    15–20% glutaraldehyde

    2% PVP or dextran, or 1% gum acacia

    20–50 mM of a suitable buffer at pH 6.8–7.4

    physiological saline to give an osmolarity of 300 mosmols without the aldehyde

    An example would be a solution prepared by dissolving 0.5 g PVP in 5 ml 0.2 M sodium cacodylate plus 2.5 ml 1.0 M NaCl and 0.5 ml 0.2 M calcium acetate or chloride, adding 10 ml of 50% glutaraldehyde, adjusting the pH with HCl if necessary and making up to a total of 25 ml with distilled water. The solution should be filtered and must be crystal clear before use.

3. Complete the perfusion with 200–250 ml of the main fixative, which is usually a more dilute glutaraldehyde solution (e.g. made up in the same way as the concentrated fixative but with 4 or 5% glutaraldehyde instead of 20%) or a combined paraformaldehyde-glutaraldehyde solution (e.g. Reese and Karnovsky 1967). The ionic composition of the main fixative should not differ significantly from the solution used in step 2.

4. Dissect out the tissues required and continue the fixation by immersion (§ 3.3.1) either in the solution used in step 3, or in an equivalent solution containing only paraformaldehyde, at 4 °C for 2–4 hr.

For enzyme cytochemistry it is advisable to omit step 2 and perhaps also to replace the saline of step 1 with a very dilute fixative about one fifth the strength of the main perfusion fluid (Reese and Karnovsky 1967). For some enzymes perfusion must be discontinued after only 2–3 min to prevent excessive inactivation (Lewis et al. 1974).

3.3.4 FIXATION OF VERY LARGE SPECIMENS

It is sometimes necessary, as in the study of surgical biopsies, to fix very large specimens. A primary fixative is required which has the maximum penetrating power, so that aldehyde fixatives containing acrolein or paraformaldehyde are chosen and trinitro compounds added (§ 2.8.4). A suitable fixative is one containing 2.5% glutaraldehyde and 2% paraformaldehyde in 0.1 M cacodylate buffer, pH 7.4, containing 0.2% trinitrocresol and 2 mM calcium chloride.

Alternatively, a rapidly penetrating fixing agent (e.g. paraformaldehyde) at high concentration (10%) is buffered with collidine (0.2 M) (§ 2.2.8) which extracts tissue components and aids penetration. Winborn and Seelig (1970) obtained good fixation in the centres of very large tissue blocks (3.5 cm^3) with such a fixative.

3.3.5 ORGAN CULTURES AND COLONIES GROWING ON AGAR

Organ cultures are very easy to fix since all that is necessary is to lift them from the medium in which they have been grown, rinse them in a balanced salt solution to remove excess medium and place them directly in the primary fixative in a small vial. They are then handled in exactly the same way as pieces of tissues fixed by immersion (§ 3.3.1), using a similar fixation schedule (§ 3.2).

Colonies of cells (tissue culture cells, bacteria, fungi, etc.) growing on agar or other solid medium are fixed by the same method (e.g. Glauert and Hopwood 1960; Tuffery 1972).

3.4 Botanical specimens

In comparison with animal tissues, botanical tissues are often believed to be relatively unaffected by removal from the plant, but ultrastructural changes are known to occur quite rapidly, mainly due to water loss. Pieces of leaves, stems or roots are therefore fixed immediately after removal from the plant. A drop of fixative is added to the required sample which is then cut, with a sharp razor blade, into thin slices (0.2–1.0 mm thick) or small cubes (about 1.0 mm^3). These pieces are then immersed in fresh fixative as quickly as possible, taking care to minimise structural damage at this stage. It is sometimes advantageous to expose tissues which float in the fixative to reduced pressure for a brief period to remove internal air and thus ensure complete contact with the fixative solution.

The early osmium tetroxide fixatives used for animal tissues gave poor results when applied to plant specimens, probably because the vacuolar contents are released at that stage in fixation when the surrounding membrane becomes permeable. The first successful preservation of plant cells was achieved using potassium permanganate (0.6–5.0%), either unbuffered or buffered with veronal–acetate, for $\frac{1}{2}$–1 hr at 2–4 °C or at 22 °C (Mollenhauer 1959). Fixation with permanganate gives excellent visualization of cellular membranes, but also causes the progressive removal of other cellular components (§ 2.7.1), and should therefore be used only for special purposes, e.g. when studying the extent of cellular membrane systems such as the endoplasmic reticulum.

Since the introduction of aldehyde fixatives, much improved preservation of cytoplasmic detail is obtained using a preliminary fixation with an aldehyde (frequently 2–6% glutaraldehyde in phosphate or cacodylate buffer, see Table 3.2), followed, after thorough rinsing, by fixation with osmium tetroxide (usually 1–2% in the same buffer). The duration of both the aldehyde and osmium tetroxide fixations may be varied and may be carried out at room temperature or in the cold. For example, a common procedure is: fixation in glutaraldehyde for $1\frac{1}{2}$–4 hr at room temperature, followed by osmium tetroxide fixation for 1–2 hr at room temperature. For some specimens the duration of fixation can be shortened considerably with equally good results (Bain and Gove 1971).

Apart from variations in the duration and temperature of fixation, there are many other procedures which are followed by different workers, such as the addition of salts or sucrose to fixatives to increase the osmolarity, or the use of mixtures of fixatives to give improved preservation (Table 3.2).

Most specimens are fixed by the simple immersion method already described, but modifications are required depending on the nature of the specimen, its size and the ease of penetration of the fixative. Small specimens, such as algal unicells or isolated chloroplasts, are fixed as a centrifuged pellet (§ 3.6.1) or in suspension (§ 3.6.2). Larger specimens, such as pieces of stem or tissues containing glands, are given a brief preliminary fixation before further dissection to obtain the required groups of cells. It is sometimes advisary to start the fixation before the tissue has been completely removed from the plant. For example, Robards (1968) irrigated the stem with cold fixative during excision in a study of differentiating xylems of willow. Some of the variations in fixatives and methods of application which have been used successfully in the study of particular specimens are indicated in Table 3.2. Since the suitability of a fixation technique can be determined only by trial and error, it is frequently advisable to compare the results of two or more procedures.

TABLE 3.2

Fixation of botanical specimens

Specimen	Fixation method	Fixing Agent	Vehicle	pH	temp.	time	Reference
Wheat coleoptiles and root tips	Immersion	6% G	0.02 M ph + $CaCl_2$	7.0	r.t.	1–2 hr	Pickett-Heaps and Northcote (1966)
Pinus cambial derivatives	Immersion	(a) 3% G	0.025 M ph	6.8–7.0	r.t. then 5°C	2 hr / 22 hr	Srivastava and O'Brien (1966)
		(b) 10% A	0.025 M ph	6.8–7.0	0°C	6 hr	
Phalaris root tips	Immersion	2% $KMnO_4$	Unbuffered		20–22°C	2 hr	López-Sáez et al. (1966)
Hibiscus pods	Immersion	3% A + 5% G	cac	?	cold	1 hr	Mollenhauer (1967)
Capsella embryos	Immersion	(a) 2% $KMnO_4$	Unbuffered		4°C	15–23 hr	Schulz and Jenzen (1968)
		(b) 6% G	0.06 M ph	6.8	4°C	4 hr	
Venus flytrap digestive gland	Immersion	(a) 2.5% G	0.02 M coll	7.4	3°C	24 hr	Scala et al. (1968)
		(b) 1% OsO_4	VeAc + 0.015 M sucrose	7.4	3°C	2.5 hr	
		(c) 2.5% G and then 3% $KMnO_4$	0.02 M coll	7.4	3°C	2 hr	
			Unbuffered		?	?	
Willow xylem	Irrigation and then immersion	(a) 6% G	0.05 M cac + 0.15 M sucrose	7.0	4°C	18 hr	Robards (1968)
		(b) 3% G	0.05 M cac + 0.15 M sucrose + salts	7.2	r.t. then 4°C	1 hr / 4 hr	

				pH	Temp.	Time	Reference
Myosurus female gametophyte	Immersion	(a) 1% KMnO₄ (b) 6% G	Unbuffered Phosphate	7.2	r.t. 4°C	2 hr} 24 hr}	Woodcock and Bell (1968)
Bean leaves	Immersion	2% G + 2% OsO₄	0.05 M cac	7.0	0–2°C	¼ hr	Falk (1969)
Tobacco pith culture	Immersion	4% P + 5% G	0.1 M ph	6.8	?	2 hr	Anderson and Cronshaw (1970)
Brown alga	?	2.5% G + 2.5% P	Cac in sea water	6.8	r.t.	1–2 hr	Bisalputra and Bisalputra (1969)
Spinach isolated chloroplasts then pellet	Suspension	1% G (final conc.) 5% G	0.05 M ph	6.8–7.0	?	2 hr	Nobel et al. (1966)
Chlorella autospores	Suspension	0.5% G + 0.4% P then 2.5% G	Diluted culture medium Diluted culture medium			1 hr 1 hr	Atkinson et al. (1972)
Fungus zoospores	Suspension then Millipore filter	1.5% G (final conc.) 1.5% G	0.005 M ph ph	6.8 6.8	22°C 22°C	10 min 1.5 hr	Hoch and Mitchell (1972)
Rice embryos	In nylon sacs weighted with glass tubing	1–6% G	0.05 M ph	7.2	4°C	12 hr	Öpik (1972)

Key: G, glutaraldehyde; A, acrolein; P, paraformaldehyde; ph, phosphate; cac, cacodylate; coll, *s*-collidine; r.t., room temperature.
Note: Aldehyde fixation is followed by fixation with osmium tetroxide.

3.5   Monolayers of cells

Cells in culture growing as a monolayer on glass, plastic or other substrate, such as a Millipore filter, are easily fixed by immersing the whole culture on its support in the fixative. It is convenient to use Columbia staining jars in which the coverslips or other supports can stand vertically and in which a number of cultures can be fixed at the same time. Alternatively, the culture is placed face-upwards in a Petri dish and is covered with a shallow drop of the fixative, care being taken to use sufficient fixative, or a drop of fixative is placed in the bottom of a Petri dish and the culture placed face-downwards on top of it. The lid is placed on the Petri dish so that the fixative does not dry out during the period of fixation and the dish is placed on a slow shaker (§ 5.2), to assist the circulation of the fixative.

Monolayers grown directly in plastic Petri dishes, or other containers, are fixed by replacing the medium with sufficient fixative to completely cover the layer of cells; about 5 ml of fixative is required for a dish 8 cm in diameter.

When studying cell surfaces it is advisable to rinse the monolayer briefly (e.g. 1 min at 37 °C) in a balanced salt solution before fixation to remove all traces of the culture medium. Otherwise components of the medium will be fixed to the surfaces of the cells.

Although this method of fixation is simple, care has to be taken in the choice of fixative and buffer since the cells are relatively unprotected against osmotic changes, as compared with cells in tissues (§ 2.2). A survey of the recent literature indicates that satisfactory preservation is usually obtained by primary fixation for 15 to 30 min at room temperature in 1–3% glutaraldehyde in 0.1 M phosphate or cacodylate buffer, pH 7.2–7.4, containing 2–3 mM calcium chloride (§ 2.4.3). Some workers prefer to use Karnovsky's (1965) paraformaldehyde–glutaraldehyde fixative (§ 2.5.5), diluted 1:2 or 1:4 with buffer. If the fixation is not satisfactory it is worthwhile testing the mixed glutaraldehyde-osmium tetroxide fixative of Hirsch and Fedorko (1968) (§ 2.8.2), fixing the monolayer for 15 to 30 min at 4 °C. In studies in which the cell surface is not being examined in detail, good fixation is obtained by fixing the cells in the presence of the growth medium. The medium is not removed and the culture is fixed by adding a volume of warm (37 °C) double-strength fixative equal in volume to the growth medium (Ross 1972).

As an alternative to immersion fixation, the monolayer can be fixed by exposing the cells to the vapours of the fixative while they are bathed in the culture medium or in a physiological salt solution (e.g. see Porter and Kallman 1953). The monolayer on its support and an open vial containing the fixative are placed together in a closed container, such as Petri dish, and left for a few minutes at room temperature.

After primary immersion or vapour fixation with an aldehyde fixative, the monolayer is rinsed in a washing solution and then fixed for 15 to 30 min at room temperature in 1% osmium tetroxide in the same buffer. If the monolayer is subsequently fixed with uranyl acetate, a period of 15 to 30 min is again sufficient. Thus the schedule for monolayers is similar to that for pieces of tissue (§ 3.2) but the time in each solution is considerably reduced.

Special problems arise during the embedding procedure, particularly when the layer of cells has to be removed from its support. Techniques for embedding monolayers are described in § 5.8.2.

3.6 Isolated cells

The fixation methods for isolated cells (cells in culture, protozoa, bacteria, blood cells etc.) are very different from the methods used for tissues. There is now no problem in obtaining rapid fixation, but it is sometimes difficult to handle the cells, particularly if only small quantities are available, and, as for monolayers of cells, the composition of the fixative has to be chosen with care.

3.6.1 FIXATION IN A PELLET

It is often possible to obtain good fixation of mammalian cells by centrifuging the cells into a pellet, carefully removing the supernatant without disturbing the pellet, and then adding the fixative. The fixative is run slowly down the side of the centrifuge tube so as to keep the pellet intact; the volume of fixative added should be at least ten times the volume of the pellet. For larger pellets, more than a few mm in depth, the cells should be resuspended in the fixative by gentle shaking to ensure that all the cells are adequately fixed. After the primary fixation the cells are then centrifuged to a pellet again before the next step in the preparative procedure. One of the fixatives used for monolayers of cells (§ 3.5) is suitable.

After the primary fixation is complete the fixative is carefully removed from the pellet and replaced with the washing solution. The cohesion of the fixed pellet depends on the type of cell, the composition of the fixative and the speed of centrifugation. Many pellets which have been fixed in glutaraldehyde are firm and show no tendency to disintegrate in the washing solution, and these can be removed from the centrifuge tube at this stage. The fact that the pellet has to be removed in this way should be borne in mind when selecting the centrifuge tube (3 ml conical tubes are suitable for processing most cells). If glass tubes are used a special set of tubes should be kept for the purpose since it is difficult to remove all traces of fixative from them. The pellet can usually be freed from the bottom of the tube fairly easily using the end of a fine spatula. If the pellet is difficult to dislodge it is preferable to use plastic tubes which can be cut down to remove the pellet.

When only a small quantity of cells is available the method developed by Malamed (1963) for cell fractions (§ 3.8) is useful (Farquhar et al. 1972).

After removal from the centrifuge tube the pellet is placed in a small drop of washing solution on a sheet of dental wax and cut into small cubes by the same method used for pieces of tissue (§ 3.3.1). Care should be taken to cut out cubes from different regions of the pellet so as to check the homogeneity of the original suspension of cells. These cubes are then passed through the subsequent steps of the preparative procedure, the fixation schedule being similar to that used for immersion fixation of cubes of tissue (§ 3.2); a second fixation for 1 hr at room temperature or 4 °C in an osmium tetroxide fixative is customary.

Pellets of fixed cells which tend to disintegrate are encapsulated by one of the methods described in § 3.6.3.

Anderson (1965) and Watanabe et al. (1967) used a modification of the method of fixing cells in a pellet as a simple technique for preparing peripheral leucocytes for electron microscopy. Anticoagulated blood is centrifuged in a horizontal type head and all but the last drop of plasma removed with a Pasteur pipette without disturbing the buffy coat or tilting the tube. A glutaraldehyde fixative is then carefully layered over the buffy coat with a pipette, using approximately the same volume of fixative as the removed plasma, and the tube is left in the vertical position for 15 min or more. The disc of fixed leucocytes, embedded in solidified plasma, can then be removed with only a thin layer of adhering erythrocytes. This disc is then placed in fresh fixative for a further period, if required, and then rinsed in several changes of buffer and cut into small cubes for further processing, care being taken to preserve the orientation of the layers.

Successful results have been obtained with this method with a variety of aldehyde fixatives; Anderson (1965) and Watanabe et al. (1967) used glutaraldehyde fixatives ranging in concentration from 2 to 6.25%, in 0.1 M phosphate or cacodylate buffer, pH 7.4, at room temperature (for 15 min) or at 4 °C (for 1 hr), while Carson et al. (1972) tested a range of paraformaldehyde fixatives of different concentrations and in different buffers, at pH 7.32 to 7.45, and found that phosphate buffers gave the most reliable and uniform fixation (Fig. 2.3). Variations in the concentration of paraformaldehyde from 0.5 to 4% and in temperature (4 or 25 °C) had little noticeable effect. All specimens were fixed for a minimum of 18 hr. The discs of blood cells were very difficult to handle because they are not cohesive after paraformaldehyde fixation.

Carson et al. (1972) also describe two other, more complicated, fixation methods for blood cells, and a simple method for bone marrow. Specimens are withdrawn from sternal punctures with a plastic syringe and left undisturbed in the syringe until clotted. The clot is removed gently and placed in a small vial containing the fixative.

3.6.2 FIXATION IN SUSPENSION IN MEDIUM

For many isolated cells good fixation is not obtained by adding the fixative to a pellet since the cells are affected by the discrepancy between the composition of the fixative and of their growth or suspending medium. Better fixation is often achieved by adding the fixative while the cells are still in suspension in medium.

The cell suspension is placed in a centrifuge tube and an equal volume (or more) of the fixative is added and mixed with the cells by gently shaking or tipping the tube. Double strength fixative can be used (e.g. Glauert and Thornley 1966) so that the final concentration of the fixative is the same as that used for pieces of tissue.

A few workers have used larger volumes of fixative than of cell suspension (e.g. Behnke 1968), while others have used small quantities of concentrated fixative (e.g. Weinbaum et al. 1970). Some typical examples of fixation of cells in suspension in medium are listed in Table 3.3 (see also Table 3.2).

The following fixation method was used successfully by Roth et al. (1970) to fix both the cell body and the axopodia of a helizoan. A glutaraldehyde fixative (12% glutaraldehyde in 0.033 M phosphate buffer, pH 5.3, 6.6 or 7.2, containing 2×10^{-5} M magnesium sulphate and 2×10^{-3} M sucrose) was added slowly to an equal volume of culture medium containing the organisms and allowed to remain for 30 sec. Then an amount of osmium tetroxide fixative equal to the glutaraldehyde fixative plus culture medium was added and the cells fixed for a further 20–30 min in the final mixed fixative, consisting of 3% glutaraldehyde and 0.5% osmium tetroxide in

TABLE 3.3

Fixation of isolated cells in suspension in medium

Type of cell	Suspending medium	Ratio medium:fixative	Fixative	pH	Temp.	Time	Typical reference
Platelets	Plasma	1:4	4% G	0.1 M cac	6.5–6.8	r.t.	1 hr	Behnke (1968)
Mast cells	Peritoneal fluid		P + G	0.1 M cac	7.4			Padawer (1968)
Leucocytes	Incubation medium	1:3	3% G	0.1 M cac or 0.1 M ph	7.4	r.t. or 4°C	2 hr overnight	Zucker-Franklin et al. (1971)
Bone marrow and blood	Plasma	1:1	0.5–4% P	0.1 M ph	7.32–7.45	4° or 25°C	2 hr to overnight	Carson et al. (1972)
Bacteria	Growth medium	10:1	1% OsO$_4$	VeAc + CaCl$_2$	6.0	r.t.	5 min	Ryter and Kellenberger (1958)
Bacteria	Growth medium	1:1	5% G	0.18 M cac	7.2	cold	1 hr	Glauert and Thornley (1966)
Bacteria	Growth medium	10:1	2% G	0.1 M cac	6.9	r.t.		Swanson et al. (1971)
Bacteria	Growth medium		concentrated G; final conc. 6% G.	0.1 M ph	7.0	4°C	overnight	Weinbaum et al. (1970)

Key: G, glutaraldehyde; P, paraformaldehyde; cac, cacodylate; ph, phosphate; VeAc, veronal-acetate.

0.016 M phosphate buffer, containing 10^{-5} M magnesium sulphate and 10^{-3} M sucrose.

Cells in suspension in medium can also be fixed by exposing them to the vapours of the fixative (Jenkins 1964) by placing two open vials, containing the cells in a drop of medium and the fixative respectively, together in a closed container, such as a Petri dish, for a few minutes.

After fixation the cell suspension is centrifuged and the pellet processed in the same way as after fixation in a pellet (§ 3.6.1).

3.6.3 ENCAPSULATING METHODS FOR PELLETS

Only certain favourable pellets are cohesive enough to remain intact after the initial fixation; others tend to disintegrate when the buffer is added for washing and material is easily lost. When this is likely to happen the pellet is 'encapsulated' by surrounding it with a soft gel of agar, fibrin or bovine serum albumin. The pellet can be encapsulated at any stage before dehydration has begun, but it is advisable for it to be done as soon as possible and thus avoiding having to centrifuge the cells to a pellet at each stage.

Agar method I

(Ryter and Kellenberger 1958)

1. Prepare a 2% solution of agar by dissolving the agar (see Appendix) in boiling distilled water or in medium. Pour the solution into a test tube while it is still molten, and place the tube in a water bath at 45 °C. At this temperature the agar just remains liquid.
2. Also place the centrifuge tube containing the pellet of fixed cells in the water bath at 45 °C.
3. Transfer a *small* drop of agar (approx. 0.03 ml) to the centrifuge tube with a *warm* pipette, and shake the tube gently to suspend the cells in the agar.
4. Immediately tilt the tube so that the agar runs down the side of the tube and forms a drop on a cool glass microscope slide.
5. After the agar has set (one or two minutes) cut the solidified agar containing the cells into small cubes, about 1 mm³, with a sharp single-edged razor blade.

These cubes are then processed in the same way as pieces of a cohesive pellet (§ 3.6.1).

It is important to ensure that the agar remains warm until the moment it is poured onto the slide or it will start to set before it is thoroughly mixed with the cells. The amount of agar added should be kept as small as possible to avoid diluting the cells more than is necessary; if too much agar is added it will be found that a thin section through the agar contains only a very few cells.

Agar method II

(Hirsch and Fedorko 1968)

1. Place the centrifuge tube containing the pellet of fixed and washed cells in a water bath at 50 °C.
2. Add a few drops of warm (50 °C) 2% agar to the pellet, using a warm pipette, and shake the tube to suspend the cells in the agar.
3. Centrifuge the tube containing the cells in agar at 750 g for 2 min in a carrier half-filled with hot tap water. The cells form a firm pellet at the bottom of the tube.
4. Cool the centrifuge tube in crushed ice to solidify the agar.
5. Add some 70% ethanol and leave the tube in ice for one hour or more.
6. Displace the pellet by carefully pipetting; remove it from the centrifuge tube and cut it into small cubes with a sharp, single-edged razor blade.

It will be noted that Hirsch and Fedorko (1968) recommend handling the agar at a higher temperature (50 °C) than that used by Ryter and Kellenberger (1958) (45 °C). There is less chance that the agar will set too soon at the higher temperature, but it is possible that the structure of the specimen will be affected, even after fixation, if the temperature rises too high. In general, it is advisable to keep the agar as cool as possible without allowing it to solidify.

Winters and Slade (1971) described a method of staining agar with Schiff's solution (leuco basic fuchsin) to make the small cubes more easily visible.

Fibrin clot method

Charret and Fauré-Fremiet (1967) showed that it is possible to use a fibrin clot instead of agar for encapsulation with the advantages that it is not necessary to raise the temperature and that the clot is more readily penetrated by dehydrating and embedding solutions than agar. The method is based on the ability of thrombin to transform a solution of fibrinogen into a fibrin clot at room temperature. Fibrin is compatible with most buffer systems, except those containing cacodylate (Sicko and Arnold 1971). By adjusting the fibrinogen content clots can be prepared with densities ranging from an open, mucus-like network (200 mg per 100 ml) to a hard, agar-like gel (1.5%). Sicko and Arnold (1971) recommend the use of freshly prepared 0.7% bovine fibrinogen for animal cell suspensions and blue green algae. Both thrombin and fibrinogen are also available commercially (see Appendix).

Furtado (1970) studied the conditions for the application of the method and developed the following technique for encapsulating pellets:

1. Dilute the commercial thrombin solution with sterile distilled water and store at a concentration of 1000 units/ml in 1 ml aliquots in vials. Dilute to a concentration of 50–100 units/ml before use.
2. Prepare a solution containing 300–350 mg fibrinogen, 160 mg sodium citrate and 850 mg sodium chloride in 100 ml distilled water.
3. Remove the washing solution from a pellet of fixed and washed cells in a centrifuge tube and add 0.2–0.3 ml of fibrinogen solution. Stir and centrifuge again to concentrate the cells, if necessary.
4. Add an equal volume of thrombin solution slowly to the supernatant fluid with a pipette. Twist a sharpened applicator stick in the fibrinogen solution until a veil-like clot becomes evident. Then twist the stick in the pellet, until the cell mass is entrapped at the tip of the stick.
5. Use the stick to transfer the clot through the subsequent steps in the preparative procedure. The clot is cut into small pieces when in 70% ethanol.

Bovine serum albumin method

Bovine serum albumin can also be used to encapsulate pellets without raising the temperature. The albumin is gelled by the addition of glutaraldehyde, and Shands (1968) recommends the following method:

1. Prepare a 2% solution of bovine serum albumin in 0.05 M Tris buffer, pH 7.5, and filter the solution through a 0.45 μm pore-size Millipore filter.
2. Resuspend the pellet of fixed cells (or other particles) in 0.25 ml of 2% bovine serum albumin in the same Tris buffer and transfer the suspension to a small cellulose centrifuge tube.
3. Add one drop of 25% (distilled) glutaraldehyde and mix it with the suspension. Centrifuge the suspension immediately in a swinging bucket type rotor. A gel forms in about 5 min and is easily detectable as the bovine serum albumin becomes opaque.
4. Remove the stopper from the tube, and cut the tube and its contents into slices about 1 mm thick. Place the slices on filter paper to remove excess moisture.
5. Cut the surrounding rims of cellulose to release the discs of gel. Dehydrate and embed these discs by standard methods.

Shands (1968) used this method successfully with macrophages and bacteria, as well as smaller particles, such as mitochondria and antigen-antibody precipitates. The only disadvantage of the method is that block staining of the pieces of gel may produce a background granularity.

3.6.4   ENCAPSULATING METHODS FOR ISOLATED CELLS

Isolated cells, such as protozoa, that are large enough to be handled individually can also be encapsulated to simplify the subsequent dehydration and embedding.

Agar method

In the technique developed by De Haller et al. (1961) a fixed cell is transferred with a pipette to a drop of 2% liquid agar on a glass slide on a hot plate at 45°C. The authors recommend that the agar be dissolved in the buffer used for the fixative. The cell is viewed under a microscope and orientated as required, while the agar is still liquid, using a fine glass rod. The glass slide is then placed in a humid atmosphere in a Petri dish and the agar is allowed to set. A small block of agar, containing the cell, is then cut out with a small scapel or sharp razor blade, and this block is then dehydrated and embedded. The cells are difficult to see in some embedding media, such as Vestopal, and De Haller et al. (1961) suggest placing a small pointer made from aluminium in the agar pointing at the cell.

Albumen method

Isolated cells can also be encapsulated in a fibrin clot using the same method as for pellets (§ 3.6.3), or in albumen, the more fluid part of natural egg white. Sawicki and Lipetz (1971) proposed the following method:

1. Prepare gelatinized glass slides by immersing them in a 0.5% aqueous solution of gelatin for 3 min and then allowing them to dry.
2. Place small drops (approx. 0.02 ml) of albumen on the slides.
3. Transfer fixed cells to the microdroplets of albumen using a micro-pipette. One or more cells may be introduced into each drop.
4. Invert the slide for about 1 min. The hanging drops elongate and the cells move to the bottom.
5. Expose the drops (still inverted) to the vapours of absolute ethanol or 40% formaldehyde in a closed dish. The drops coagulate in about 5 min and become opaque.
6. Preparations coagulated by formaldehyde are washed in water and then placed in 70% ethanol; the others are transferred directly to 70% ethanol. Stand the slides vertically in the ethanol in a Coplin jar.

7. Dehydrate the drops in increasing concentrations of ethanol, and remove the drops from the slides and place them in small vials before impregnation with the embedding medium. The hardened drops are easily removed from the gelatinized slide.

3.6.5 COLLECTING CELLS ON MILLIPORE FILTERS

A convenient method of collecting isolated cells for fixation is to use a Millipore membrane filter. Filters with various pore sizes are available (see Appendix), 0.45 μm being suitable for cell cultures and some bacteria, 0.22 μm for smaller bacteria and bacterial spheroplasts and 50 nm for viruses. Cells are collected either by filtering the suspension through the Millipore membrane in the usual way or a sterile filter is placed in a Petri dish and the concentrated suspension placed gently in the centre. The filter is then incubated at 37 °C for 2–3 hr to allow the cells to attach firmly (McCombs et al. 1968). The cells on the filter are fixed and dehydrated in the same way as monolayers of cells grown on filters or other substrata (§ 3.5), the filter being cut into pieces before impregnation with embedding medium. Propylene oxide may dissolve the filter and so xylene or toluene is used instead (see § 5.8.2d).

3.7 Cell fractions

Subcellular fractions, such as microsomes, mitochondria or bacterial cell walls, are handled in the same way as isolated cells, but higher speeds of centrifugation are usually required because of the smaller size of the particles. Adequate fixation is often obtained by centrifuging the fraction to a pellet and then adding the fixative as described in § 3.6.1. Plastic centrifuge tubes should be used for easy removal of the fixed pellet. Some membrane-bounded particles, such as mitochondria which are sensitive to the osmotic conditions, show signs of damage when fixed in this way, and it is then advisable to fix the particles while they are still in suspension in the isolation medium. The fraction is fixed by mixing it with an equal volume (or more) of a fixative, which contains the same concentration of sucrose as the suspending medium, as in the method for suspensions of cells (§ 3.6.2). Alternatively, sufficient purified, concentrated glutaraldehyde is added to the suspension to give a final concentration of glutaraldehyde of about 1%.

Millonig and Marinozzi (1968) have shown that glutaraldehyde is preferable to osmium tetroxide for the fixation of fractions suspended in sucrose because sucrose inhibits proper fixation of proteins by osmium tetroxide. They also recommend first fixing pellets from cell fractions in glutaraldehyde vapour for several minutes before adding the liquid fixative. The composition of the glutaraldehyde fixative does not appear to be critical, concentrations of glutaraldehyde ranging from 0.5–3.0% in 0.1 M cacodylate buffer being customary, calcium chloride (2–3 mM) being added for the fixation of membraneous particles. Fractions in the form of a pellet are fixed for 30 min to 1 hr, the time depending on the size of the pellet, while much shorter times (e.g. 10 min) are required for fixation in suspension. These fractions are centrifuged to a pellet after fixation, before washing and post-fixation with osmium tetroxide. If the pellet tends to disintegrate in the washing solution, it is encapsulated by one of the methods described for pellets of isolated cells (§ 3.6.3).

The pellet is removed from the centrifuge tube and cut into small cubes or strips for post-fixation in osmium tetroxide, care being taken that samples are taken from all levels in the pellet so that the heterogeneity of the sample is fully explored. Whenever possible, it is advisable to cut the pellet into strips aligned in the direction of the axis of the centrifuge tube and then to section these strips longitudinally. In this way the different layers of the pellet are visible in the final section.

Baudhuin et al. (1967) found that the heterogeneity in some pellets does not occur solely in the direction of the centrifugal force, but also extends along the surface of the preparation. As a result particles of different types occur with differing frequencies in the centre and in the periphery of the preparation. To overcome this problem they developed a method of preparing very thin (about 10 µm) pellicles of packed particles in which heterogeneity is solely in a direction perpendicular to the surface of the pellicle. The method involves filtration on a Millipore filter under pressure and is somewhat complicated. It has the great advantage that sections covering the whole depth of the pellicle can be photographed in a single field in the electron microscope.

3.8   Methods for very small quantities of material

When only small quantities of cells or cell fractions are available it is necessary to use correspondingly small centrifuge tubes in order to obtain a compact pellet that is not spread in a thin layer over the bottom of the tube. Malamed (1963) suggested using disposable, polyethylene tubes of 0.4 ml capacity such as those that fit into the high speed Microfuge (Beckman Instruments, Spinco Division, see Appendix). A small pellet is transferred on the tip of a

Pasteur pipette to a microcentrifuge tube about two-thirds filled with fixative. After the required fixation time, a pellet is formed by centrifugation (Fig. 3.3) and the fixative is withdrawn. The pellet is released by cutting the tube transversely just above the top of the pellet with a sharp razor blade, making a similar cut just beneath the pellet and then dislodging the pellet by poking a wooden toothpick into the lower, smaller hole. The pellet is then handled in the same way as larger pellets (§ 3.6.1).

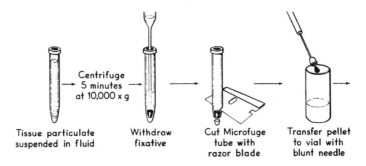

Fig. 3.3. Schematic representation of the microcentrifugal method for processing material for electron microscopy (from Malamed 1963).

Another method of handling very small quantities of material is described by Marikovsky and Danon (1963, 1966). The fraction is placed in a polyethylene capillary tube (1 mm internal diameter, 70 mm long) containing a column of phthalate esters. The capillary tube is sealed at both ends and then centrifuged in a microhematocrit centrifuge at 12,000 *g* for 15 min. The fraction separates into different layers and the required layers are obtained by cutting the tube into segments with a razor blade. These layers are fixed, washed and dehydrated, the fraction being expelled from the capillary tube in 70% ethanol (or acetone) using a thin wire.

A third method of handling very small amounts of material is to carry out the whole preparative procedure, from fixation to embedding, in a pyramidal-tipped plastic capsule (BEEM, see Appendix) (§ 5.2). The material is centrifuged to a pellet for each change of solution and care has to be taken to resuspend the sample properly, particularly during infiltration with the embedding medium which should be chosen to have low viscosity (e.g. Spurr epoxy resin) (§ 5.3.2a). Demaree (1970) suggests tying a thread to the capsule so that it can be easily lowered into a tapered centrifuge tube, the thread being taped to the outside of the tube. Alternatively, a special carrier for holding a number of capsules can be used. Kondo and Takemura (1965) describe a holder made out of cork, while Blackburn (1968) recommends a carrier made from brass. Centrifuge tubes modified to hold BEEM capsules are available commercially (see Appendix).

3.9   Special fixation schedules and methods

The majority of specimens are well fixed by the methods already described with a standard fixation schedule, similar to those outlined in § 3.2. Certain specimens or chemical components require the application of modified fixation methods, and some of these are outlined briefly in this section.

3.9.1   LIPIDS

Lipids are poorly preserved by routine fixation methods (Stein and Stein 1971) so that a high proportion of the lipid may be lost during the subsequent dehydration (Chapter 4). Alterations in the dehydration procedure are the best way to preserve lipids, but various modifications to the standard fixation procedure of glutaraldehyde followed by osmium tetroxide help to reduce the amount of lipid lost, including:

 i. Addition of calcium chloride to the glutaraldehyde, and possibly also the subsequent osmium tetroxide, fixative to reduce the extraction of *phospholipids* (e.g. Mitchell 1969).

 ii. Addition of digitonin to the primary aldehyde, or aldehyde plus osmium tetroxide, fixative to reduce extraction of *free cholesterol* (§ 2.8.5).

 iii. Fixation by Baker's (1946) formal-calcium procedure, as modified by Casley-Smith (1963). Specimens are fixed firstly with 4% formaldehyde plus 1% calcium chloride, in 0.03 M veronal–acetate buffer, pH 7.4, containing 7.5% sucrose, for 18 hr at 3 °C, and then secondly in 5% potassium dichromate plus 1% calcium chloride for 18 hr at 20 °C and then for 24 hr at 60 °C. This fixation method improves the preservation of *butter* and *maize oil* (Casley-Smith 1963). The preservation of *phospholipids* is no better than after routine fixation in glutaraldehyde followed by osmium tetroxide (Cope and Williams 1969).

 iv. Tricomplex fixation (see § 2.8.3) for the preservation of *labile lipids*.

3.9.2   CELLS WITH THICK WALLS

Thick-walled cells, such as yeasts, and certain other specimens, such as sea urchin eggs, are difficult to fix well because the fixative does not penetrate easily. Various techniques have been suggested for aiding the penetration of the fixative, including:

i. Use of a fixative containing acrolein, the most rapidly penetrating of the aldehyde fixatives (§ 2.6.1), or potassium permanganate (§ 2.7.1), (e.g. Wallace et al. 1968).

ii. Mechanical breakage of the cell before or during fixation by cutting (Bluemink 1970) or by shaking with glass beads. For example, Damsky et al. (1969) poured a yeast culture over ice into flasks containing sufficient paraformaldehyde or glutaraldehyde to give a final concentration of fixative of 1 or 2.5% respectively, and after 20 min or less ruptured the cells in a Braun glass bead mill. Small pieces of the cell wall were removed but the cell outlines were not destroyed. The broken cells were fixed for a further period in the aldehyde fixative and then with osmium tetroxide.

3.9.3 BACTERIA

It is difficult to obtain good fixation of bacteria by primary fixation with osmium tetroxide because of the labile nature of the nuclear material, which is associated with relatively little protein and which is not surrounded by a nuclear membrane. In consequence Ryter and Kellenberger (1958) developed a special method for the fixation of bacteria (in the period before the introduction of aldehyde fixatives), and this schedule is still widely used. The method was designed with the specific aim of stabilising the nuclear material.

Ryter-Kellenberger standard fixation for bacteria

1. *Pre-fixation.* Mix 30 ml of bacteria suspended in tryptone culture medium (1% bacto-tryptone, Difco, and 0.5% sodium chloride) in a centrifuge tube with 3 ml of Ryter-Kellenberger veronal-acetate buffered osmium tetroxide fixative (§ 2.3.5), and centrifuge immediately for 5 min at 1800 g.

2. *Main fixation.* Resuspend the pellet in 1.0 ml of Ryter-Kellenberger fixative and 0.1 ml of tryptone medium, and leave overnight (about 16 hr) at room temperature.

3. Dilute the suspension with 8 ml of Ryter-Kellenberger veronal-acetate buffer (§ 2.2.7) and centrifuge for 5 min at 1800 g.

4. Resuspend the pellet in about 0.03 ml of warm agar and put the drop on a microscope slide (§ 3.6.3). Cut the hardened clot into small cubes, approx. 1 mm³.

5. Place the cubes in 0.5% uranyl acetate in Ryter-Kellenberger veronal-acetate buffer for 2 hr at room temperature. Dehydrate and embed.

This fixation schedule preserves the nuclear material of bacteria in a finely fibrillar state, and also results in a clear demonstration of membranes, due to the presence of calcium and uranyl ions which help to preserve lipids (Silva et al. 1968; Burdett and Rogers 1970). Recent studies suggest that 'pre-fixation' in the presence of a low concentration of osmium tetroxide may lead to a reorganisation of the membranes of mesosomes (Silva 1971). In general, this step is omitted, the pellet of bacteria, or cells on Millipore filters or agar, being placed directly in full-strength fixative.

Improved results are also obtained by fixing the bacteria first in a glutaraldehyde fixative, either in suspension or in a pellet, using a similar fixation schedule as for mammalian cells. This primary glutaraldehyde fixation is followed by fixation for 1-16 hr in Ryter-Kellenberger osmium tetroxide fixative and then by fixation in uranyl acetate before dehydration and embedding.

3.9.4 MARINE ORGANISMS

Sea water is added to fixatives for marine organisms. Osmium tetroxide is dissolved directly in sea water (e.g. Afzelius 1956), while glutaraldehyde fixatives are prepared by mixing double-strength glutaraldehyde (e.g. 6%) with an equal volume of sea water, or by buffering the fixative with sodium cacodylate dissolved in filtered sea water (e.g. Anderson and Personne 1970).

Potassium permanganate fixatives have also been used for the fixation of marine organisms. Doggenweiler and Heuser (1967) recommend a 1.5% solution of potassium permanganate in artificial sea water for the fixation (for 3 hr at 4°C) of the prawn nerve sheath, while Martin and Rosenberg (1968) used 0.6% potassium permanganate in Tris-buffered sea water at pH 7.5 to 8.0 for the fixation (for ½ to 2 hr) of squid giant axon fibres.

3.9.5 ORGANISMS LIVING IN EXTREME CONDITIONS

Uni- or multi-cellular organisms which live in environments of very low or very high ionic strength, or at alkaline or acid pH, are very difficult to fix. It is paramount that the fixative has a pH near neutrality and that the concentration of electrolytes matches that of the interior of the cell (Millonig 1973).

3.9.6 SPECIAL FIXATION SCHEDULES

The standard fixation schedules outlined in § 3.2 are suitable for the great majority of specimens. Some modified schedules for special purposes have been described in this chapter, and these are listed in Table 3.4, together with a selection of fixation schedules which are not widely used but may prove to be of value in special applications.

TABLE 3.4

Fixation schedules

Sequence of fixatives	Type of specimen	Typical reference
Standard fixation schedules		
1. Aldehydes + Ca^{2+}; OsO_4	Various	Karnovsky (1965)
2. Aldehydes; OsO_4; UA	Various	Farquhar and Palade (1965)
Fixation schedules for special purposes		
3. F + Ca^{2+}; $K_2Cr_2O_7$ + Ca^{2+}	Lipids	Casley-Smith (1963)
4. G + OsO_4; UA	Isolated cells and monolayers	Hirsch and Fedorko (1968)
5. OsO_4 + $K_2Cr_2O_7$	Various	Dalton (1955)
6. OsO_4; UA	Bacteria	Ryter and Kellenberger (1958)
7. $KMnO_4$	Plants	Mollenhauer (1959)
8a. OsO_4 in sea water	Marine	Afzelius (1956)
8b. G in sea water	organisms	Anderson and Personne (1970)
8c. $KMnO_4$ in sea water	Marine organisms	Doggenweiler and Heuser (1967)
Fixation schedules not widely used		
9. G or F; $K_2Cr_2O_7$; OsO_4	Isolated cells	Sugihara et al. (1966)
10. G; $Pb(NO_3)_2$ + $K_3Fe(CN)_6$; OsO_4	Lung lipids	Dermer (1969)
11. G; $KMnO_4$	Plants	Scala et al. (1968)
12. G + A; $KMnO_4$	Plants	Hayat (1970)
13. Aldehyde; OsO_4 + $K_3Fe(CN)_6$	Various	Karnovsky (1971)
14. G + OsO_4; OsO_4	Various	Franke et al. (1969)
15. G + OsO_4; G	Amphibian	Bluemink (1970)
16. G + F + trinitro compounds	Steroid-secreting tissues	Ito and Karnovsky (1968)
17. OsO_4; F	Connective tissues	Ross and Klebanoff (1969)
18. OsO_4; G; UA	Brain	Kanaseki and Kadota (1969)
19. OsO_4 + $K_2Cr_2O_7$; UA	Tissue culture	Zeigel et al. (1969)
20. $KMnO_4$; $K_2Cr_2O_7$ + UN	Yeast	Wallace et al. (1968)
21. $KMnO_4$ + Ca^{2+}; $K_2Cr_2O_7$	Yeast	Kawakami (1961)

Key: UA, uranyl acetate; F, formaldehyde; G, glutaraldehyde; A, acrolein; UN, uranyl nitrate.

REFERENCES

Afzelius, B. A. (1956), The ultrastructure of the cortical granules and their products in the sea urchin egg as studied with the electron microscope, Expl Cell Res. *10*, 257.

Anderson, D. R. (1965), A method of preparing peripheral leucocytes for electron microscopy, J. Ultrastruct. Res. *13*, 263.

Anderson, R. and J. Cronshaw (1970), Sieve element pores in *Nicotiana* pith culture, J. Ultrastruct. Res. *32*, 458.

Anderson, W. A. and P. Personne (1970), The localization of glycogen in the spermatozoa of various invertebrate and vertebrate species, J. Cell Biol. *44*, 29.

Atkinson, A. W., B. E. S. Gunning and P. C. L. John (1972), Sporopollenin in the cell wall of *Chlorella* and other algae: ultrastructure, chemistry and incorporation of [14C] acetate, studied in synchronous culture, Planta *107*, 1.

Bain, J. M. and D. W. Gove (1971), Rapid preparation of plant tissues for electron microscopy, J. Microscopy *93*, 159.

Baker, J. R. (1946), The histochemical recognition of lipine, Q. Jl Microsc. Sci. *87*, 441.

Baldwin, K. M. (1970), The fine structure and electrophysiology of heart muscle cell injury, J. Cell Biol. *46*, 455.

Baudhuin, P., P. Evrard and J. Berthet (1967), Electron microscopic examination of subcellular fractions. 1. The preparation of representative samples from suspensions of particles, J. Cell Biol. *32*, 181.

Behnke, O. (1968), Electron microscopical observations on the surface coating of human blood platelets, J. Ultrastruct. Res. *24*, 51.

Bisalputra, T. and A. A. Bisalputra (1969), The ultrastructure of chloroplast of a brown alga *Sphacelaria* sp. 1. Plastid DNA configuration – the chloroplast genophore, J. Ultrastruct. Res. *29*, 151.

Blackburn, W. R. (1968), Carrier centrifuge tube for gelatin and polyethylene capsules, J. Ultrastruct. Res. *23*, 362.

Bluemink, J. G. (1970), The first cleavage of the amphibian egg, J. Ultrastruct. Res. *32*, 142.

Bodian, D. (1970), An electron microscopic characterization of classes of synaptic vesicles by means of controlled aldehyde fixation, J. Cell. Biol. *44*, 115.

Bohmann, S. O. and A. B. Maunsbach (1970), Effects on tissue fine structure of variations in colloid osmotic pressure of glutaraldehyde fixatives, J. Ultrastruct. Res. *30*, 195.

Brightman, M. W. and T. S. Reese (1969), Junctions between intimately apposed cell membranes in the vertebrate brain, J. Cell Biol. *40*, 648.

Burdett, I. D. J. and H. J. Rogers (1970), Modification of the appearance of mesosomes in sections of *Bacillus licheniformis* according to the fixation procedures, J. Ultrastruct. Res. *30*, 354.

Carson, F., J. A. Lynn and J. H. Martin (1972), Ultrastructural effect of various buffers, osmolality, and temperature on paraformaldehyde fixation of the formed elements of blood and bone marrow, Tex. Rep. Biol. Med. *30*, 125.

Casley-Smith, J. R. (1963), Some observations on the fixation and staining of lipids, Jl R. microsc. Soc. *81*, 235.

Charret, R. and E. Fauré-Fremiet (1967), Technique de rassemblement de microorganismes: préinclusion dans un caillot de fibrine, J. Microscopie *6*, 1063.

Cope, G. H. and M. A. Williams (1969), Quantitative studies on the preservation of choline and ethanolamine phosphatides during tissue preparation for electron microscopy. II. Other preparative methods, J. Microscopy *90*, 47.

Dalton, A. J. (1955), A chrome–osmium fixative for electron microscopy, Anat. Rec. *121*, 281.

Damsky, C. H., W. M. Nelson and A. Claude (1969), Mitochondria in anaerobically-grown lipid-limited brewer's yeast, J. Cell Biol. *43*, 174.

De Haller, G., C. F. Ehret and R. Naef (1961), Technique d'inclusion et d'ultramicrotomie, destinée à l'étude du développement des organelles dans une cellule isolée, Experientia *17*, 524.

Demaree, R. S. (1970), Isolation of avian trypanosomes by selective centrifugation and subsequent processing for electron microscopy, Stain technol. *45*, 304.

Dermer, G. B. (1969), The fixation of pulmonary surfactant for electron microscopy. 1. The alveolar surface lining layer, J. Ultrastruct. Res. *27*, 88.

Doggenweiler, C. F. and J. E. Heuser (1967), Ultrastructure of the prawn nerve sheath. Role of fixative and osmotic pressure in vesiculation of thin cytoplasmic laminae, J. Cell Biol. *34*, 407.

Falk, H. (1969), Rough thylakoids: polysomes attached to chloroplast membranes, J. Cell Biol. *42*, 582.

Farquhar, M. G., D. F. Bainton, M. Baggiolini and C. De Duve (1972), Cytochemical localization of acid phosphatase activity in granule fractions from rabbit polymorphonuclear leukocytes, J. Cell Biol. *54*, 141.

Farquhar, M. G. and G. E. Palade (1965), Cell junctions in amphibian skin, J. Cell Biol. *26*, 263.

Franke, W. W., S. Krein and R. M. Brown (1969), Simultaneous glutaraldehyde-osmium tetroxide fixation with postosmication. An improved fixation for electron microscopy of plant and animal cells, Histochemie *19*, 162.

Furtado, J. S. (1970), The fibrin clot: a medium for supporting loose cells and delicate structures during processing for microscopy, Stain technol. *45*, 19.

Gardner, H. A., G. Simon and M. D. Silver (1969), The fine structure of rabbit platelets fixed in acetaldehyde, Anat. Rec. *163*, 509.

Gil, J. and E. R. Weibel (1969–70), Improvements in demonstration of lining layer of lung alveoli by electron microscopy, Respiration Physiol. *8*, 13.

Glauert, A. M. and D. A. Hopwood (1960), The fine structure of *Streptomyces coelicolor*. 1. The cytoplasmic membrane system, J. biophys. biochem. Cytol. *7*, 479.

Glauert, A. M. and M. J. Thornley (1966), Glutaraldehyde fixation of Gram-negative bacteria, Jl R. microsc. Soc. *85*, 449.

Hayat, M. E. (1970), Principles and techniques of electron microscopy, Vol. 1, Biological applications (Van Nostrand Reinhold, New York).

Hirsch, J. G. and M. E. Fedorko (1968), Ultrastructure of human leukocytes after simultaneous fixation with glutaraldehyde and osmium tetroxide and 'postfixation' in uranyl acetate, J. Cell Biol. *38*, 615.

Hoch, H. C. and J. E. Mitchell (1972), The ultrastructure of zoospores of *Aphanomyces euteiches* and of their encystment and subsequent germination, Protoplasma *75*, 113.

Ito, S. (1962), Post-mortem changes of the plasma membrane, Proc. 5th Int. Congr. Electron Microscopy, Philadelphia *2*, L-5.

Ito, S. and M. J. Karnovsky (1968), Formaldehyde-glutaraldehyde fixatives containing trinitro compounds, J. Cell Biol. *39*, 168a.

Ito, S. and R. J. Winchester (1963), The fine structure of the gastric mucosa in the bat, J. Cell Biol. *16*, 541.

Jenkins, R. A. (1964), Osmium tetroxide vapor fixation of protozoan cells, J. Cell Biol. *23*, 46A.

Kanaseki, T. and K. Kadota (1969), The 'vesicle in a basket', J. Cell Biol. *42*, 202.

Karnovsky, M. J. (1965), A formaldehyde–glutaraldehyde fixative of high osmolality for use in electron microscopy, J. Cell Biol. *27*, 137A.

Karnovsky, M. J. (1967), The ultrastructural basis of capillary permeability studied with peroxidase as a tracer, J. Cell Biol. *35*, 213.

Karnovsky, M. J. (1971), Use of ferrocyanide-reduced osmium tetroxide in electron microscopy, Proc. 14th Ann. Meeting Amer. Soc. Cell Biol., p. 146.

Kawakami, N. (1961), A permanganate–chrome fixative and lead acetate staining for electron microscopy of microorganisms, J. Electron Microscopy *10*, 14.

Kondo, K. and K. Takemura (1965), Application of epoxy resin embedding method for free cells, J. Electron Microscopy *14*, 50.

Lewis, P. R., D. P. Knight and M. A. Williams (1974), Staining methods for thin sections, in: Practical methods in electron microscopy, A. M. Glauert, ed. (North-Holland, Amsterdam).

López-Sáez, J. F., G. Giménez-Martin and M. C. Risueño (1966), Fine structure of the plasmodesm, Protoplasma *61*, 81.

Lumb, W. V. (1963), Small animal anesthesia (Lea and Febiger, Philadelphia).

Malamed, S. (1963), Use of a microcentrifuge for preparation of isolated mitochondria and cell suspensions for electron microscopy, J. Cell Biol. *18*, 696.

Marikovsky, Y. and D. Danon (1963), A micromethod for fixation and embedding of separated blood cell fractions for electron microscopy, J. Ultrastruct. Res. *18*, 176.

Marikovsky, Y. and D. Danon (1966), A micromethod for fixation and embedding of blood cells separated according to their density, Proc. 6th Int. Congr. Electron Microscopy, Kyoto *2*, 31.

Marks, I. and L. G. Briarty (1970), A syringe-filter for processing specimens through fluids used in plastic embedding, Stain technol. *45*, 36.

Martin, R. and P. Rosenberg (1968), Fine structural alterations associated with venom action on squid giant axon fibers, J. Cell Biol. *36*, 341.

Maunsbach, A. B. (1966), Absorption of I^{125}-labeled homologous albumin of rat kidney proximal tubule cells, J. Ultrastruct. Res. *15*, 197.

Maunsbach, A. B., S. C. Madden and H. Latta (1962), Variations in fine structure of renal tubular epithelium under different conditions of fixation, J. Ultrastruct. Res. *6*, 511.

McCombs, R. M., M. Benyesh-Melnick and J. P. Brunschwig (1968), The use of Millipore filters in ultrastructural studies of cell cultures and viruses, J. Cell Biol. *36*, 231.

Millonig, G. (1973), Personal communication, Laboratory of Molecular Embryology, University of Naples.

Millonig, G. and V. Marinozzi (1968), Fixation and embedding in electron microscopy, in: Adv. Optical and Electron Microscopy, Vol. 2, R. Barer and V. E. Cosslett, eds. (Academic Press, New York and London), p. 251.

Mitchell, C. D. (1969), Preservation of the lipids of the human erythrocyte stroma during fixation and dehydration for electron microscopy, J. Cell Biol. *40*, 869.

Mollenhauer, H. H. (1959), Permanganate fixation of plant cells, J. biophys. biochem. Cytol. *6*, 431.

Mollenhauer, H. H. (1967). The fine structure of mucilage secreting cells of *Hibiscus esculentus* pods, Protoplasma *63*, 353.

Nobel, P. S., S. Murakami and A. Takamiya (1966), Localization of light-induced strontium accumulation in spinach chloroplasts, Plant and Cell Physiol. *7*, 263.

Öpik, H. (1972), Some observations on coleoptile cell ultrastructure in ungerminated grains of rice (*Oryza sativa* L), Planta *102*, 61.

Padawer, J. (1968), Uptake of colloidal thorium dioxide by mast cells, J. Cell Biol. *40*, 747.

Palade, G. E. (1956), The fixation of tissues for electron microscopy, Proc. 3rd Int. Congr. Electron Microscopy, London, p. 129.

Palfrey, A. J. and D. V. Davies (1966), The fine structure of chondrocytes, J. Anat. *100*, 213.

Pease, D. C. (1964), Histological techniques for electron microscopy, 2nd edition (Academic Press, New York and London).

Peters, A., C. C. Proskauer and I. R. Kaiserman-Abramof (1968), The small pyramidal neuron of the rat cerebral cortex, J. Cell Biol. *39*, 604.

Pickett-Heaps, J. D. and D. H. Northcote (1966), Organization of microtubules and endoplasmic reticulum during mitosis and cytokenesis in wheat meristems, J. Cell Sci. *1*, 109.

Porter, K. R. and F. Kallman (1953), The properties and effects of osmium tetroxide as a tissue fixative with special reference to its use for electron microscopy, Expl Cell Res. *4*, 127.

Reese, T. S. and M. J. Karnovsky (1967), Fine structural localization of a blood-brain barrier to exogenous peroxidase, J. Cell Biol. *34*, 207.

Rhodin, J. (1954), Correlation of ultrastructural organization and function in normal and experimentally changed proximal convoluted tubule cells of the mouse kidney, Thesis, Stockholm.

Robards, A. W. (1968), On the ultrastructure of differentiating secondary xylem in willow, Protoplasm *65*, 449.

Robertson, J. D. (1963), The occurrence of a subunit pattern in the unit membranes of club endings in Maunther cell synapses in goldfish brain, J. Cell Biol. *19*, 201.

Ross, R. (1972), Personal communication, Department of Pathology, University of Washington, Seattle.

Ross, R. and S. J. Klebanoff (1967), Fine structural changes in uterine smooth muscle and fibroblasts in response to estrogen, J. Cell Biol. *32*, 155.

Roth, L. E., D. J. Pihlaja and Y. Shigenaka (1970), Microtubules in the helizoan axopodium, J. Ultrastruct. Res. *30*, 7.

Ryter, A. and E. Kellenberger (1958), Etude au microscope électronique de plasmas contenant de l'acide désoxyribonucléique, Z. Naturf. *13*, 597.

Sawicki, W. and J. Lipetz (1971), Albumen embedding and individual mounting of one or many mammalian ova on slides for fluid processing, Stain technol. *46*, 261.

Scala, J., D. Schwab and E. Simmons (1968), The fine structure of the digestive gland of Venus's flytrap, Am. J. Bot. *55*, 649.

Schulz, R. and W. Jensen (1968), *Capsella* embryogenesis: the early embryo, J. Ultrastruct. Res. *22*, 376.

Schultz, R. L. and N. M. Case (1970), A modified aldehyde perfusion technique for preventing certain artifacts in electron microscopy of the central nervous system, J. Microscopy *92*, 69.

Shands, J. W. (1968), Embedding free-floating cells and microscopic particles: serum albumin coagulum – epoxy resin, Stain technol. *43*, 15.

Sicko, L. M. and W. J. Arnold (1971), Fibrin clots for electron microscopy and cytochemistry of minute biological samples, Proc. 29th Ann. Meeting EMSA, p. 472.

Silva, M. T. (1971), Changes induced in the ultrastructure of the cytoplasmic and intracytoplasmic membranes of several Gram-positive bacteria by variations in OsO_4 fixation, J. Microscopy *93*, 227.

Silva, M. T., F. C. Guerra and M. M. Magalhães (1968), The fixative action of uranyl acetate in electron microscopy, Experientia *24*, 1074.

Sjöstrand, F. S. and L.-G. Elfvin (1962), The layered asymmetric structure of the plasma membrane in the exocrine pancreas cells of the cat, J. Ultrastruct. Res. *7*, 504.

Smith, R. E. and M. G. Farquhar (1966), Lysosome function in the regulation of the secretory process in cells of the anterior pituitary gland, J. Cell Biol. *31*, 319.

Srivastava, L. M. and T. P. O'Brien (1966), On the ultrastructure of cambium and its vascular derivatives, I. Cambium of *Pinus strobus* L, Protoplasma *61*, 257.

Stein, O. and Y. Stein (1971), Light and electron microscopic radioautography of lipids: techniques and biological applications, Adv. Lipid Res. *9*, 1.

Sugihara, R., K. J. Lee, T. Sugioka and G. Yasuzumi (1966), Experiments on triple fixation for electron microscopy, Proc. 6th Int. Congr. Electron Microscopy, Kyoto *2*, 25.

Swanson, J., S. J. Kraus and E. C. Gotschlich (1971), Studies on gonococcus infections, J. exp. Med. *134*, 886.

Tormey, J. McD. (1963), Fine structure of the ciliary epithelium of the rabbit, with particular reference to 'infolded membranes', 'vesicles', and the effects of Diamox, J. Cell Biol. *17*, 641.

Trump, B. F., P. J. Goldblatt and R. E. Stowell (1962), An electron microscopic study of early cytoplasmic alteration in hepatic parenchymal cells of mouse liver during necrosis *in vitro* (autolysis), Lab. Invest. *11*, 986.

Tuffery, A. A. (1972), Surface extensions on BHK cells grown in monolayers and agar suspension, J. Cell Sci. *10*, 123.

Wallace, P. G., M. Huang and A. W. Linnane (1968), The biogenesis of mitochondria, J. Cell Biol. *37*, 207.

Watanabe, I., S. Donahue and N. Hoggatt (1967), Method for electron microscopic studies of circulating human leukocytes and observations on their fine structure, J. Ultrastruct. Res. *20*, 366.

Weinbaum, G., D. A. Fischman and S. Okuda (1970), Membrane modifications in nutritionally induced filamentous *Escherichia coli* B, J. Cell Biol. *45*, 493.

Winborn, W. B. and L. L. Seelig (1970), Paraformaldehyde and *s*-collidine – a fixative for preserving large blocks of tissue for electron microscopy, Tex. Rep. Biol. Med. *28*, 347.

Winters, C. and M. Slade (1971), Embedding free-floating cells in stained agar for rapid screening and subsequent ultramicrotomy, Stain technol. *46*, 161.

Woodcock, C. L. F. and P. R. Bell (1968), Features of the ultrastructure of the female gametophyte of *Myosurus minimus*, J. Ultrastruct. Res. *22*, 546.

Zeigel, R. F., G. Rabotti and M. V. A. Smith (1969), Electron microscopic observations on the association of viruses with membrane systems in hamster tumor cells propagated in tissue culture, J. natn. Cancer Inst. *43*, 653.

Zucker-Franklin, D., P. Elsbach and E. J. Simon (1971), The effect of the morphine analog levorphanol on phagocytosing leukocytes, Lab. Invest. *25*, 415.

Chapter 4

Dehydration

Most embedding media are not soluble in water and consequently fixed specimens are 'dehydrated' by passing them through a sequence of solutions, the last of which is miscible with the embedding medium. The two most widely used dehydrating agents are ethanol and acetone; they give very similar results with the majority of specimens and ethanol is usually preferred for general use since acetone readily takes up water and incomplete dehydration may result.

Polyester resins are not soluble in ethanol and specimens are usually dehydrated in acetone, or they are dehydrated in ethanol and then immersed in carefully dried acetone. Styrene can also be used as an intermediate solvent for polyester resins (e.g. Kurtz 1961).

Most epoxy resins are soluble in ethanol and acetone, but they mix much more readily with propylene oxide (1,2–epoxy propane). Consequently propylene oxide is frequently used as the last stage of dehydration before embedding in an epoxy resin. Xylene, toluene and styrene are also suitable as intermediate solvents for most epoxy resins, but propylene oxide is to be preferred because it is a reactive diluent and will not separate from the epoxy resin during polymerization (Luft 1961). Consequently no problems arise if a small amount of propylene oxide remains in the embedding medium. Propylene oxide should not be used if the specimen has been stained with phosphotungstic acid (PTA) during dehydration since propylene oxide extracts PTA.

4.1 Chemical and morphological effects of dehydration

The major chemical effect of dehydration is the extraction of lipids from the specimen by the dehydrating agents (ethanol, acetone, propylene oxide etc). which are excellent lipid solvents. As much as 95% of the lipid in the specimen may be lost (Korn and Weisman 1966), the majority being extracted at concentrations of ethanol or acetone in water higher than 70% and in propylene oxide. Full details of lipid extraction during dehydration are given in the excellent review by Stein and Stein (1971). For practical purposes it can be concluded that lipid extraction can be reduced by:

(i) Fixing the specimen with osmium tetroxide before dehydration (e.g. Cope and Williams 1969a).

(ii) Fixing the specimen with uranyl acetate before dehydration (see § 2.9).

(iii) Using acetone in preference to ethanol as the dehydrating agent (e.g. Ashworth et al. 1966).

(iv) Dehydrating the specimens as rapidly as possible at low temperature.

(v) Using the partial dehydration schedule of Idelman (1964, 1965) (see § 4.2.3).

Proteins may also be extracted during dehydration. For example, Luft and Wood (1963) found that about 4% of tissue proteins were extracted during dehydration following osmium tetroxide fixation. In contrast to lipids, the proteins were extracted in the lower concentrations of ethanol in water, none being lost in 95% or 100% ethanol or in propylene oxide.

The extraction of lipid and protein during dehydration is accompanied by shrinkage of tissues, cells and subcellular components (Fig. 1.1). This extraction leads to a decrease in the diameter of sea urchin eggs (Kushida 1962), a decrease in the periodicity of nerve myelin (Moretz et al. 1969), and the shrinkage of chromosomes with an accompanying production of abrupt bends in spindle microtubules (Jensen and Bajer 1969). Some striated muscle filaments shorten by as much as 10% during dehydration in ethanol, although there is little change in dimensions during dehydration in acetone (Page and Huxley 1963). Acetone also seems to be preferable to ethanol for the dehydration of brain tissue, since specialised cell contacts in which surface membranes are closely apposed appear to be better preserved (Johnston and Roots 1967) (Fig. 4.1). However, Schultz and Karlsson (1972) consider that it is impossible to show which dehydrating agent best preserves the membrane relationships existing *in vivo*.

In lung tissue no significant change in volume is observed if the dehydration is started in 70% ethanol in water (Weibel and Knight 1964).

4.2   Dehydration schedules

Dehydration is accomplished by passing the fixed specimen through a graded series of solutions of increasing concentration of the dehydrating agent in water, ending with pure (or absolute) dehydrating agent. The specimens remain in the small vials in which they were fixed (Fig. 3.2) throughout dehydration. One solution is removed carefully with a Pasteur pipette and the next poured on. Care should be taken to ensure that the volume of solution used is much greater (at least 10 times) than the volume of the specimen.

For general morphological studies the effects produced during dehydration are slight compared with the changes occurring during fixation, and the schedule chosen for dehydration is not critically important. Specimens may be dehydrated at room temperature, or in the cold. If dehydration is started in the cold, the temperature is raised to room temperature when the specimens are in a 90% or 95% solution of the dehydrating agent in water.

Fixed specimens are sometimes washed in buffer before dehydration to remove excess fixative, but this is usually not necessary.

4.2.1 STANDARD DEHYDRATION SCHEDULE

A typical dehydration schedule for small blocks of tissue, 0.5 mm or less on each side, is:

50% ethanol or acetone in water	10 min
70% ethanol or acetone in water	10 min
95% ethanol or acetone in water	10 min
100% ethanol or acetone	15 min
100% ethanol or acetone	15 min

Specimens to be embedded in an epoxy resin are then passed through two changes of 15 min each of the intermediate solvent, propylene oxide, xylene or toluene. Propylene oxide must be flushed down the sink with a large volume of water after use, and care must be taken not to inhale the fumes.

These times are probably the maximum required, even for compact, well-fixed tissues. It is often unnecessary to use concentrations of dehydrating agent less than 70% and so the stage in 50% ethanol or 50% acetone can be omitted. Specimens that have to be stored overnight during dehydration are kept in 70% ethanol or 70% acetone in the cold. There is very little loss of lipid during long periods of storage in 70% ethanol (Idelman 1965).

4.2.2 RAPID DEHYDRATION

When required the dehydration can be made much more rapid. Dehydration times can be decreased by:

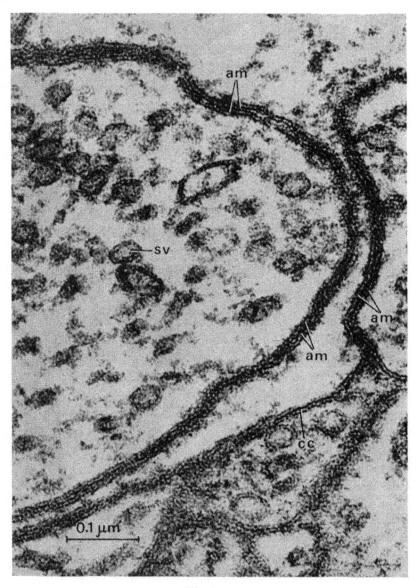

Fig. 4.1a

Fig. 4.1.   Effects of dehydration. Comparison of the appearance of closely apposed plasma membranes in rat brain after dehydration in ethanol or acetone. (a) Acetone dehydration: the plasma membranes are very closely apposed (am) or completely fused (cc). (b) Ethanol dehydration: adjacent plasma membranes (um) are separated by an extracellular space (es) about 10 nm wide. (m) Mitochondrion. (sv) Synaptic vesicle (from Johnston and Roots 1967).

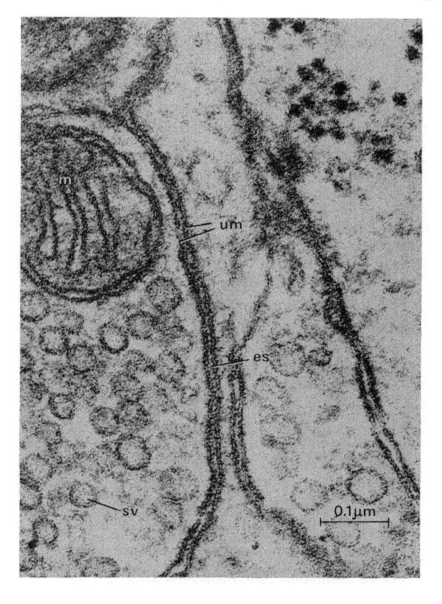

Fig. 4.1b

(i) agitating the specimens throughout dehydration

(ii) carrying out the dehydration at room temperature (and not in the cold)

(iii) using more changes of each solution

(iv) using smaller blocks of tissue.

A typical rapid dehydration schedule suggested by Coulter (1967) has proved successful with a variety of specimens. Small blocks of fixed tissue, 0.25 mm^3 or $2 \times 2 \times 0.1$ mm are passed through a series of solutions of increasing ethanol concentration to absolute ethanol in 2 min with constant agitation. Three changes of absolute ethanol are made in 1 min, and then three changes of propylene oxide at two minute intervals. The subsequent infiltration with embedding medium can also be done rapidly if the specimens are continuously agitated and a vacuum is applied (see § 5.3.4c).

Comparisons of rapid and standard dehydration schedules have not shown any obvious differences in the preservation of fine structure (Tormey 1965; Bencosme and Tsutsumi 1970).

4.2.3 PARTIAL DEHYDRATION

Idelman (1964, 1965) proposed a dehydration schedule which omits the use of concentrations of ethanol higher than 70% in water and of propylene oxide, thus avoiding the stages of dehydration during which the majority of the extraction of lipids occurs (Stein and Stein 1971). The dehydration schedule also depends on the fact that the epoxy resin Epon 812 (§ 5.3) is soluble in 70% ethanol in water.

Partial dehydration schedule I

(Idelman 1964)

Wash in buffer	brief	r.t.
30% ethanol in water	15 min	r.t.
70% ethanol in water	30 min	r.t.
70% ethanol in water	30 min	r.t.
1:1 mixture of 70% ethanol and Epon		
embedding mixture (§ 5.3.1.)	30 min	r.t.
and then	30 min	37°C
Epon embedding mixture	1 hr	37°C
Infiltrate and embed in Epon (§ 5.3.4a)		

Key: r.t., room temperature.

A modified version of this partial dehydration schedule has been used extensively by Stein and Stein (1971) for a great variety of lipids. They found that the loss of some lipids is reduced even further by decreasing the temperature to 0 °C.

Partial dehydration schedule II

(Stein and Stein 1967)

70% ethanol in water	5 min	0 °C
70% ethanol in water	5 min	0 °C
95% ethanol in water	5 min	0 °C
95% ethanol in water	5 min	0 °C
Epon 812	1 hr	0 °C
Epon 812	1 hr	0 °C
Epon 812	1 hr	0 °C
Epon embedding mixture (§ 5.3.1.)	overnight	4 °C

Infiltrate and embed in Epon

For thin pellets of cells and for monolayers of cells a rapid partial dehydration schedule can be used.

Rapid partial dehydration schedule

(Robbins and Jentzsch 1967)

20% ethanol–80% water	30 sec	r.t.
60% ethanol–40% Epon 812	5 min	r.t.
30% ethanol–70% Epon 812	5 min	r.t.
100% Epon 812	5 min	r.t.

Infiltrate and embed in Epon embedding mixture

Key: r.t., room temperature

Robbins and Jentzsch (1967) also developed a rapid method for embedding (§ 5.3.4c)

4.3 Dehydrating agents

Ethanol and acetone are the standard dehydrating agents for the great majority of procedures, with the addition of propylene oxide or styrene as intermediate solvents before embedding in epoxy or polyester resins respectively. Various alternative dehydrating agents have been tested, and these are considered in this section. It is also possible to use water-soluble embedding media for dehydration as discussed in the following Chapter (§ 5.6).

4.3.1 ETHYLENE GLYCOL

Pease (1966a) investigated the possibility of dehydrating unfixed specimens using increasing concentrations of 'inert', water-soluble, permeable, organic substances as a technique for the preservation of proteins in their native state. The success of the method is thought to depend upon adequate stabilisation and consequent immobilisation of macromolecular systems before physiological damage is produced in cytomembranes. Once this immobilisation is achieved, the final stages of dehydration can be completed in a variety of ways. The method is considered to be a physical method of preservation, comparable with freeze-substitution (see § 6.3).

The simplest procedure involves the use of the dehydrating agent ethylene glycol, which affects the native conformation of proteins less than most other organic solvents (Sjöstrand and Barajas 1968). Thin slices of tissue (less than 0.5 mm thick) are isolated in a balanced salt solution containing 10% ethylene glycol, and then dehydrated as rapidly as possible to avoid excessive shrinkage. The slices are placed in a watch glass in 5 ml of salt solution containing 10% ethylene glycol and then the concentration of ethylene glycol is increased to over 50% in about 3 min by adding 7 ml of ethylene glycol drop by drop while the watch glass is agitated vigorously. At a concentration of 50% ethylene glycol the tissue becomes transparent and the subsequent dehydration schedule is not important. Dehydration is usually completed in a further 2 to 3 min. Specimens can be stored indefinitely in pure ethylene glycol at 4 °C with no obvious deterioration.

After dehydration specimens are embedded directly in hydroxypropyl methacrylate (§ 5.6.2b), or in an epoxy or polyester resin. Pease (1966a) originally suggested that Cellosolve (the monoethyl ester of ethylene glycol) be used as an intermediate solvent before embedding in an epoxy or polyester resin, but Sjöstrand and Barajas (1968) found that this resulted in complete destruction of the tissue. Consequently propylene oxide and/or toluene should be used (Pease 1973).

The main damage produced by this dehydration procedure is the extraction of lipids during dehydration and during infiltration with the embedding medium (Cope and Williams 1968, 1969b). This damage can be reduced by fixing the specimens in glutaraldehyde, or in glutaraldehyde and osmium tetroxide, in 70% ethylene glycol (Pease 1966a). Alternatively specimens are fixed briefly ($\frac{1}{2}$ to $1\frac{1}{2}$ min) with glutaraldehyde before dehydration in ethylene glycol (Sjöstrand and Barajas 1968).

After dehydration in ethylene glycol and embedding in an epoxy or polyester resin, the blocks can be sectioned relatively easily onto a trough containing ethylene glycol using a diamond knife, the whole process being anhydrous. Hydroxypropyl methacrylate blocks can only be sectioned with difficulty (Pease 1966b).

The results of this technique are of interest for comparison with conventionally-dehydrated material, but characteristic artefacts result from the tensions set up during dehydration (Pease 1973). The structure of the tissue in a surface layer about 0.1 mm thick, and of all the tissue below a depth of about 0.5 mm, is destroyed. The tissue in the narrow optimally-preserved zone shows evidence of considerable shrinkage; nuclear outlines appear scalloped and mitochondria are irregular in shape. However, much of the finer structure of the tissue is retained and some components, such as polysaccharides, are preserved better than by conventional dehydration techniques (Pease 1973).

Ethylene glycol is also useful as an intermediate solvent for specimens which are naturally dehydrated. For example, Morris (1968) fixed slices of dehydrated cysts of *Artemia salina* in dry osmium tetroxide vapours, 'washed' them in two changes of ethylene glycol and then embedded them in Epon.

4.3.2 POLYETHYLENE GLYCOL

Polyethylene glycol 200 (Carbowax 200) is miscible with both water and epoxy resins and is thus suitable as a dehydrating agent (Kushida 1963). Polyethylene glycol has a higher viscosity than ethanol and acetone and so longer soaking times are required. Fixed specimens are dehydrated in 50%, 70% and 90% polyethylene glycol in water for 20–40 min each, and finally in 2 changes of absolute polyethylene glycol for 40–80 min. Specimens are immersed in a 1:1 mixture of polyethylene glycol and epoxy resin embedding mixture for 1 hr before infiltration and embedding in the epoxy resin.

Polyethylene glycol is not readily soluble in polyester resins and so methyl methacrylate is used as an intermediate solvent (Kushida and Fujita 1970).

4.3.3 PROPYLENE OXIDE

Propylene oxide (1,2–epoxy propane) is used routinely between ethanol dehydration and infiltration with epoxy resin embedding media. Kushida (1961) showed that propylene oxide can also be used as the sole dehydrating agent for specimens to be infiltrated with Epon 812. Propylene oxide is not readily miscible with water at concentrations above 40% (Spurr 1969) and the dehydration schedule relies on the fact that water is appreciably miscible with Epon 812.

Propylene oxide dehydration schedule

(Kushida 1961)

20% propylene oxide in water	10–15 min	r.t.
40% propylene oxide in water	10–15 min	r.t.
1:1 mixture of 70% p.o. and Epon	10–15 min	r.t.
1:1 mixture of 90% p.o. and Epon	10–15 min	r.t.
1:1 mixture of 100% p.o. and Epon	10–15 min	r.t.
Epon 812	30 min	40°C
Epon 812	30 min	40°C

Key: p.o., propylene oxide; r.t., room temperature.

The incomplete miscibility of propylene oxide with water at higher concentrations can be overcome by adding a small amount of ethanol (2.5 ml of ethanol to each 100 ml of propylene oxide). A standard dehydration schedule can then be used (Spurr 1969).

4.3.4   OTHER DEHYDRATING AGENTS

Acid alcohol was used as a dehydrating agent, following osmium tetroxide fixation, by Sandborn (1962) and he reported good preservation of matrix material and membranes. Casley-Smith (1967) found that acid alcohol did not help to preserve lipids in general, although he thought that it may improve the preservation of phospholipids.

Spurr (1969) has used hexylene glycol as a dehydrating agent with the aim of helping to avoid loss of lipids.

Watson (1962) used a mixture of acrolein (§ 2.6.1) and ethanol as a dehydrating agent for the preservation of proteins, while Robison and Lipton (1969) used acrolein alone, following fixation in a mixture of acrolein and potassium dichromate (§ 2.8.6). This procedure revealed subunit detail in microtubules which appeared 'negatively' stained.

REFERENCES

Ashworth, C. T., J. S. Leonard, E. H. Eigenbrodt and F. J. Wrightman (1966), Hepatic intracellular osmiophilic droplets. Effect of lipid solvents during tissue preparation, J. Cell Biol. *31*, 301.
Bencosme, S. A. and V. Tsutsumi (1970), A fast method for processing biologic material for electron microscopy, Lab. Invest. *23*, 447.

Casley-Smith, J. R. (1967), Some observations on the electron microscopy of lipids, Jl R. microsc. Soc. *87*, 463.

Cope, G. H. and M. A. Williams (1968), Quantitative studies on neutral lipid preservation in electron microscopy, Jl R. microsc. Soc. *88*, 259.

Cope, G. H. and M. A. Williams (1969a), Quantitative studies on the preservation of choline and ethanolamine phosphatides during tissue preparation for electron microscopy. I. Glutaraldehyde, osmium tetroxide, Araldite methods, J. Microscopy *90*, 31.

Cope, G. H. and M. A. Williams (1969b), Quantitative studies on the preservation of choline and ethanolamine phosphatides during tissue preparation for electron microscopy. II. Other preparative methods, J. Microscopy *90*, 47.

Coulter, H. D. (1967), Rapid and improved methods for embedding biological tissues in Epon 812 and Araldite 502, J. Ultrastruct. Res. *20*, 346.

Idelman, S. (1964), Modification de la technique de Luft en vue de la conservation des lipides en microscopie électronique, J. Microscopie *3*, 715.

Idelman, S. (1965), Conservation des lipides par les techniques utilisées en microscopie électronique, Histochemie *5*, 18.

Jensen, C. and A. Bajer (1969), Effects of dehydration on the microtubules of the mitotic spindle, J. Ultrastruct. Res. *26*, 367.

Johnston, P. V. and B. I. Roots (1967), Fixation of the central nervous system by perfusion with aldehydes and its effect on the extracellular space as seen by electron microscopy, J. Cell Sci. *2*, 377.

Korn, E. D. and R. A. Weisman (1966), Loss of lipids during preparation of amoebae for electron microscopy, Biochim. biophys. Acta *116*, 309.

Kurtz, S. M. (1961), A new method for embedding tissues in Vestopal W, J. Ultrastruct. Res. *5*, 468.

Kushida, H. (1961), Propylene oxide as a dehydrating agent for embedding with epoxy resins, J. Electron Microscopy, *10*, 203.

Kushida, H. (1962), A study of cellular swelling and shrinkage during fixation, dehydration and embedding in various standard media, J. Electron Microscopy, *11*, 135.

Kushida, H. (1963), An improved epoxy resin 'Epok 533', and polyethylene glycol 200 as a dehydrating agent, J. Electron Microscopy, *12*, 167.

Kushida, H. and K. Fujita (1970), Polyethylene glycol 200 as a dehydrating agent for embedding with polyester resins, J. Electron Microscopy, *19*, 391.

Luft, J. H. (1961), Improvements in epoxy resin embedding methods, J. biophys. biochem. Cytol. *9*, 409.

Luft, J. H. and R. L. Wood (1963), The extraction of tissue protein during and after fixation with osmium tetroxide in various buffer systems, J. Cell Biol. *19*, 46A.

Moretz, R. C., C. K. Akers and D. F. Parsons (1969), Use of small angle X-ray diffraction to investigate disordering of membranes during preparation for electron microscopy. I. Osmium tetroxide and potassium permanganate, Biochem. biophys. Acta *193*, 1.

Morris, J. E. (1968), Dehydrated cysts of *Artemia salina* prepared for electron microscopy by totally anhydrous techniques, J. Ultrastruct. Res. *25*, 64.

Page, S. G. and H. E. Huxley (1963), Filament lengths in striated muscle, J. Cell Biol. *19*, 369.

Pease, D. C. (1966a), The preservation of unfixed cytological detail by dehydration with 'inert' agents, J. Ultrastruct. Res. *14*, 356.

Pease, D. C. (1966b), Anhydrous ultrathin sectioning and staining for electron microscopy, J. Ultrastruct. Res. *14*, 379.

Pease, D. C. (1973), Substitution techniques, in: Advanced techniques in biological electron microscopy, J. K. Koehler, ed. (Springer-Verlag, Berlin). p. 35.

Robbins, E. and G. Jentzsch (1967), Rapid embedding of cell culture monolayers and suspensions for electron microscopy, J. Histochem. Cytochem. *15*, 181.

Robison, W. G. and B. H. Lipton (1969), Advantages of dichromate-acrolein fixation for preservation of ultrastructural detail, J. Cell Biol. *43*, 117a.

Sandborn, E. (1962), A method of dehydration for improved visualization of lipids, membranes and other cytoplasmic inclusions in tissues to be embedded in Epon, Proc. 5th Int. Congr. Electron Microscopy, Philadelphia *2*, P-12.

Schultz, R. L. and U. L. Karlsson (1972), Brain extracellular space and membrane morphology variations with preparative procedures, J. Cell Sci. *10*, 181.

Sjöstrand, F. S. and L. Barajas (1968), Effect of modifications in conformation of protein molecules on structure of mitochondrial membranes, J. Ultrastruct. Res. *25*, 121.

Spurr, A. R. (1969), A low-viscosity epoxy resin embedding medium for electron microscopy, J. Ultrastruct. Res. *26*, 31.

Stein, O. and Y. Stein (1967), Lipid synthesis, intracellular transport, storage, and secretion. 1. Electron microscope radioautographic study of liver after injection of tritiated palmitate or glycerol in fasted and ethanol-treated rats, J. Cell Biol. *33*, 319.

Stein, O. and Y. Stein (1971), Light and electron microscopic radioautography of lipids: techniques and biological applications, Adv. Lipid Res. *9*, 1.

Tormey, J. McD. (1965), Artifactual localization of ferritin in the ciliary epithelium *in vitro*, J. Cell Biol. *25*, 1.

Watson, M. L. (1962), Considerations of nucleic acid morphology in fixed tissues, Proc. 5th Int. Congr. Electron Microscopy, Philadelphia *2*, 0-5.

Weibel, E. R. and B. W. Knight (1964), A morphometric study on the thickness of the pulmonary air–blood barrier, J. Cell Biol. *21*, 367.

Chapter 5

Embedding

In the final stage of preparing a biological specimen in a form suitable for thin-sectioning it is infiltrated with a liquid embedding medium which is then polymerized to produce a solid block.

5.1 Embedding media for electron microscopy

Ideally an embedding medium should have the following properties:
  (i) Easy availability
 (ii) Uniformity from one batch to another
(iii) Solubility in dehydrating agents
(iv) Low viscosity as a monomer
 (v) Polymerize uniformly
(vi) Little volume change on polymerization
(vii) Ease of sectioning
(viii) Stability in the electron beam

Three main types of embedding media are in general use; the epoxy resins, the polyester resins and the methacrylates. None of the embedding media so far developed possesses all the desired qualities, and some of the properties may be incompatible with others. For example, a medium which polymerizes with little shrinkage usually has a high initial viscosity. The embedding media widely used in electron microscopy are listed in Table 5.1, and their properties are discussed in the following sections of this chapter. The epoxy resins (§ 5.3) are the most widely used since they have most of the properties required. The polyester resins (§ 5.4) have similar excellent properties, but are less easily available and some of the components of the embedding medium are not stable during storage. The methacrylates (§ 5.5) polymerize unevenly and are unstable under electron bombardment. Consequently, they are not suitable for general use, although the water-soluble methacrylates (§ 5.6.2) are valuable as embedding media in cytochemical studies.

TABLE 5.1

Embedding media for electron microscopy

Manufacturer	Trade name	Introduced by:
Epoxy Resins		
Shell	Epon 812 (originally 562) and Epon 815	Kushida (1959)
	Epon 812 (named Epikote 812 in Europe)	Gibbons (1959); Finck (1960); Luft (1961)
Ciba	Araldite CY 212 (originally Araldite M) (sold by Fluka and others as Durcupan ACM)	Glauert et al. (1956); Glauert and Glauert (1958)
	Araldite 6005	Mollenhauer (1959)
	X133/2097 (sold by Fluka as Durcupan)	Stäubli (1960)
	Araldite 502	Finck (1960); Luft (1961)
	Araldite 506	Mollenhauer (1964)
Marblette	Maraglas 655	Freeman and Spurlock (1962); Spurlock et al. (1963)
Oken Shoji	Epok 533	Kushida (1963b)
Dow	DER 332	Lockwood (1964)
	DER 334	Winborn (1965)
Union Carbide	ERL 4206	Spurr (1969)
Polyester resins		
Martin Jaeger	Vestopal W	Ryter and Kellenberger (1958a, b).
Riken Goseijushi	Rigolac 2004 and Rigolac 70F	Kushida (1960a)
Pittsburg Plate Glass	Selectron 5003 and Selectron 5214	Low and Clevenger (1962)
Société des Usines Chimiques Rhône-Poulenc	Rhodester 1108	Argagnon and Enjalbert (1964)
B.I.P. Chemicals	Beetle 4116 and Beetle 4134	Rampley and Morris (1972)
Methacrylates		
Various	*n*-butyl methacrylate and methyl methacrylate	Newman et al. (1949)
Various	glycol methacrylate	Rosenberg et al. (1960)
Various	hydroxypropyl methacrylate	Leduc and Holt (1965)

5.1.1 CHEMICAL AND MORPHOLOGICAL EFFECTS OF EMBEDDING MEDIA

The changes occurring in the specimen during embedding are slight compared with those occurring during fixation (Chapter 2) and dehydration (Chapter 4). There is little dimensional change during infiltration of the embedding medium into most specimens (Fig. 1.1; Kushida 1962b; Moretz et al. 1969a), although isolated cells, such as amoebae, may shrink in viscous embedding media (Szubinska 1971). There is also little shrinkage (usually less than 2%) during the polymerization of epoxy and polyester resins (Fig. 1.1). Large dimensional changes result from embedding in methacrylates, including three-dimensional methacrylate (Fig. 1.1). Epoxy resin monomers extract small amounts of lipid, the amount depending on the preceding fixation and dehydration schedule (Korn and Weisman 1966; Casley-Smith 1967; Cope and Williams 1968; 1969a, b; Stein and Stein 1971). The methacrylates are powerful lipid solvents, even at low temperatures (Cope and Williams 1968; 1969b; Stein and Stein 1971).

5.2 Standard embedding methods

Embedding media should be handled with great care. Many of the components can cause dermatitis if they come into contact with the skin, and consequently all procedures should be carried out as cleanly as possible. In addition, the vapours of the components are often toxic and particular care should be taken when handling embedding media of low viscosity which tend to be more volatile. Embedding media should be handled in a fume cupboard whenever possible.

The standard method for embedding specimens in gelatin or polyethylene capsules is similar for all types of embedding medium. After fixation (Chapter 3) and dehydration (Chapter 4) the specimens will be soaking in the final dehydrating solution in small vials (Fig. 3.2) and are ready for infiltration with the embedding medium. Complete embedding mixtures, containing the accelerator or activator, are usually used for infiltration and are prepared freshly during the later stages of dehydration.

The specimens are infiltrated with the embedding medium by passing them through a sequence of solutions until the dehydrating agent has been completely replaced by the final embedding mixture. The different schedules required for the various embedding media are described in the later sections of this chapter. In a typical schedule, the dehydrating agent (or intermediate solvent) is removed with a pipette and replaced with a solution consisting of equal parts of the dehydrating agent and the embedding medium. After 1 hr this solution is replaced with pure embedding medium and the vials are left overnight with their tops off to allow any remaining dehydrating agent to evaporate. It is difficult to pour viscous embedding media and they are best transferred in a wide-mouthed pipette.

The infiltration of specimens with viscous embedding media can be speeded up by placing the vials on a slow rotary shaker. Jurand and Ireland (1965) and Kushida (1969a) have designed rotary mixers specially for embedding (Figs. 5.1 and 5.2), which are available commercially (see Appendix). These mixers can also be used to shorten fixation and dehydration schedules, and in the preparation of fixatives and embedding mixtures (Fig. 5.3).

As an alternative to using a shaker, a continuous stream of gas bubbles can be used to keep the embedding medium circulating around the specimens (Kushida and Fujita 1971). Nitrogen is used for methacrylates, styrene and polyester resins (since oxygen inhibits their polymerization), and dehydrated air for epoxy resins.

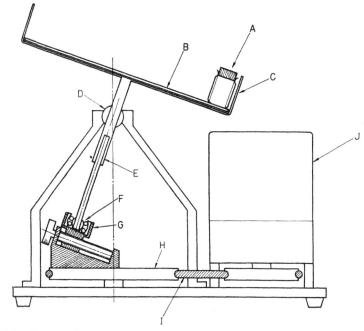

Fig. 5.1.   Cross-sectional diagram of the rotary shaker designed by Kushida (1969a): (A) vial; (B) metal plate; (C) pan; (D) spherical bearing; (E) shaft; (F) self-aligning bearing; (G) carriage (H) pulley; (I) belt; (J) induction motor.

Fig. 5.2. A general view of the shaker designed by Kushida (1969a).

Delicate tissue specimens may be adversely affected by sudden changes in viscosity and it is therefore advisable to increase the concentration of the embedding medium in a continuous process. Steinbrecht and Ernst (1967) describe a method of continuous infiltration in which the vials containing the specimens in propylene oxide are placed on a rotating mixer similar to the device designed by Jurand and Ireland (1965) and the embedding medium (an epoxy resin) is added dropwise.

In the standard embedding method, gelatin or polyethylene capsules are used as embedding moulds to provide blocks of a convenient size for thin sectioning. Gelatin capsules are obtainable in a range of sizes (No. 00, 0, 1, 2, etc.) (See Appendix); size No. 0 is convenient for general use. Polyethylene capsules (Fig. 5.4) of various designs are available commercially and are known as BEEM or TAAB capsules, depending on the supplier (see Appendix). These capsules have pyramid-shaped ends and yield pre-formed blocks that require the minimum of trimming before sectioning (Reid 1974).

Fig. 5.3.   The shaker designed by Kushida (1969a) adapted to hold beakers and flasks for the preparation of fixatives and embedding media.

Taab also supply capsules made of polypropylene for polymerization at high temperatures.

The capsules must be completely dry before use and are best stored in a 60 °C oven overnight, or longer, before use. A set of capsules is placed in a suitable support, such as the lid of a cardboard box punched with an array of holes (Fig. 5.5) or special trays which are available commercially (see Appendix). A specimen is removed from the vial with fine tweezers or a pipette and placed at the bottom of the capsule. The capsule is filled about half full with the final embedding mixture and the specimen is guided back to the exact bottom of the capsule, if necessary. A label indicating the experiment and sample number (use a typewriter or Indian ink) is placed in each capsule in such a position that the number is visible from outside the capsule (Fig. 5.5).

The embedding medium is polymerized by heating the capsules in an oven or by irradiation with ultra-violet (UV) light. Polymerization by UV irradiation has the advantage that it can be carried out at room temperature or in the cold. For example, Shinagawa and Uchida (1961) polymerized methacrylate by UV light in a cold room and found that the temperature of the specimens did not rise above 24 °C.

An ordinary sun lamp is adequate as a source of UV light (e.g. Leduc and Holt 1965). The capsules are placed a few inches away from the lamp.

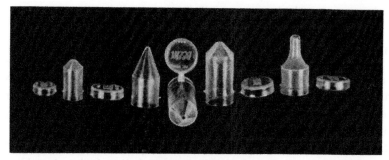

Fig. 5.4. Various designs of BEEM polyethylene capsules designed to produce pre-shaped blocks.

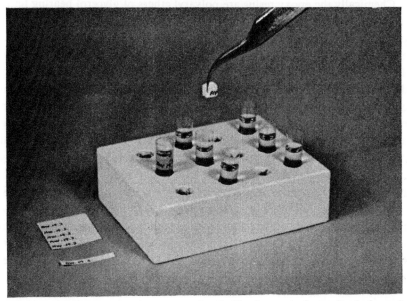

Fig. 5.5. A cardboard lid makes a convenient holder for a set of capsules during polymerization of the embedding medium. A label is about to be inserted into one of the gelatin capsules.

Alternatively, a special apparatus can be constructed, such as that designed by Shinagawa et al. (1962) in which 5 UV lamps are attached to the inside of a metal arch. Cole (1968) described a simple and cheap apparatus which is easily constructed in the laboratory. Polymerization was carried out at 4 °C, 30 °C or 42 °C by placing the apparatus in a freezer at − 16 °C, or in a constant temperature room at 27 °C or in an incubator at 35 °C respectively. The UV source was a General Electric black-light lamp. A more complex apparatus is described by McGee-Russell and De Bruijn (1964). Suitable UV lamps are available commercially (see Appendix).

5.3   Epoxy resins

The epoxy resins were tested as embedding media for electron microscopy (Maaløe and Birch-Andersen 1956; Glauert et al. 1956) in a search for an alternative to the methacrylates (§ 5.5). In common with polyester resins (§ 5.4), epoxy resins have the advantages that they polymerize uniformly with very little change of volume and are relatively stable in the electron beam (Reid 1974). Their main disadvantage is their high viscosity which necessitates a lengthy infiltration procedure as compared with the methacrylates.

Chemically the epoxy resins are polyaryl ethers of glycerol bearing terminal epoxy groups. They range from viscous liquids to fusible solids and may be polymerized by a variety of bifunctional setting agents which add across the epoxy groups of the resin molecules to give three-dimensional structures. These setting agents are commonly aliphatic polyamines or aromatic anhydrides, but the number of possible setting agents is very large and this enables a wide range of products with very different characteristics to be obtained. The polymerization can be made to occur with a volume shrinkage as low as 2%, a comparable figure for methacrylate being 15 to 20%. This relatively small change in volume is due, on the one hand, to the polymerization being an addition process, and, on the other, to the fact that the resin is highly associated in the uncured state (Glauert and Glauert 1958).

The uniformity of the polymerization of epoxy resins is due to the fact that the hardening is an addition reaction and is not dependent on an initiator. In consequence the reaction is not affected by the presence of impurities, unlike the methacrylates in which components of the specimen itself may initiate polymerization (§ 5.5).

5.3.1   EPOXY RESIN EMBEDDING MEDIA

The first epoxy resin embedding medium to be tested (Maaløe and Birch-

Andersen 1956) consisted of a Shell epoxy resin (designated EPO) and an aliphatic polyamine hardener (diethylene triamine). This hardener is very viscous and the mixture sets rapidly even at room temperature. The mixture also has a limited compatibility with ethanol and it is very difficult to get adequate penetration of the embedding medium into the specimen. Subsequently Glauert et al. (1956) tested a series of Ciba epoxy resins (Araldites) and an anhydride hardener (dodecenyl succinic anhydride, DDSA). The reaction between the epoxy resin and this hardener is much slower than in an EPO mixture and the hardener is less viscous than diethylene triamine. In addition, the mixture of Araldite and DDSA is readily soluble in ethanol so that good penetration of the embedding medium into the specimen is easily achieved. The reaction between the epoxy resin and DDSA is slow and an amine accelerator is added to hasten polymerization.

The Araldites (502, 506 and 6005) manufactured in the USA are not identical to those made in England (CY 212) and some difficulties were experienced in using them as embedding media (Richardson et al. 1960). Consequently Finck (1960) and Luft (1961) investigated the suitability of Shell epoxy resins (Epons, named Epikotes in Europe) for electron microscopy, and showed that embedding media based on Epon 812 are excellent for the purpose. At the same time Kushida (1959) in Japan developed an embedding medium containing a mixture of Epon 812 and Epon 815, and later he used an epoxy resin (Epok) manufactured in Japan by Oken Shoji (Kushida 1963b). Embedding media based on epoxy resins manufactured by Marblette (Maraglas), Dow (DER) and Union Carbide (ERL) have also been successfully used for electron microscopy.

Most of the epoxy resin embedding media have the same basic components as the original Araldite mixture, consisting of an epoxy resin, a hardener and an accelerator, sometimes with the addition of a plasticizer or epoxy flexibilizer. These embedding media are listed in Table 5.2. It will be noted that some media contain more than one epoxy resin (sometimes from more than one manufacturer) and that others contain more than one hardener. In some mixtures a reactive epoxy flexibilizer (Cardolite NC-513, DER 732 or DER 736) acts as hardener as well as plasticizer. The final blocks range from crystal clear (e.g. Maraglas 655) to light gold (e.g. Araldite) in colour.

All the embedding media listed in Table 5.2 are suitable for electron microscopy and a decision on which to use is usually based on availability and the uniformity of the epoxy resin from one batch to another. There are reports that Maraglas 655 (Erlandson 1964) and DER 334 (Winborn 1965) show some variability and that it is desirable to run blanks on each new can.

TABLE 5.2
Epoxy resin embedding media

Epoxy resin		Hardener	Accelerator	Additives	Features	Reference
Araldite CY 212	10 ml	DDSA 10 ml	DMP-30 0.5 ml	DBP 1 ml	First use of Araldite	Glauert and Glauert (1958)
Araldite CY 212	10 ml	DDSA 10 ml	DMP-30 0.5 ml		Plasticizer omitted	Richardson et al. (1960)
Araldite CY 212	10 ml	DDSA 9 ml + MNA 1 ml	DMP-30 0.4 ml		For harder specimens	Robson (1964)
Araldite CY 212	10 ml	DDSA 10 ml	BDMA 0.4 ml		Standard 1974	Parker (1972)
Araldite 502	10 ml	DDSA 10 ml	BDMA 0.2 ml		U.S. equivalents of original Araldite mixture	Finck (1960)
Araldite 502	27 ml	DDSA 23 ml	DMP-30 0.75–1.0 ml			Luft (1961)
Araldite 502	68 ml	DDSA 19 ml	DMP-30 3 ml	TAC 10 ml	TAC added to improve sectioning properties	Winborn (1965) (Table 5.10)
Araldite 502	10 ml	DDSA 7.8 ml	DMP-30 1.5%		Anhydride/epoxy ratio adjusted	Coulter (1967) (Table 5.9)
Araldite 6005	10 ml	DDSA 7 ml	DMP-30 0.24 ml		U.S. equivalents of original Araldite mixture	Richardson et al. (1960)
Araldite 6005	10 ml	DDSA 10 ml	BDMA 0.2–0.3 ml	DBP 1 ml		Parsons and Darden (1961)
Araldite 506	50 ml		BDMA 3%	Cardolite NC-513 + DBP 25 ml 1–2 ml	DDSA replaced with reactive epoxy flexibilizer	Mollenhauer (1964) (Table 5.8)
Epon 812	10 ml	DDSA 19 ml	BDMA 1%		First use of Epon 812 in the U.S.	Finck (1960)
Epon 812	10 ml	DDSA 18 ml + HHPA 1 ml	BDMA 1%		Add HHPA for harder block	Finck (1960)

Resin	Amount	Hardener	Accelerator	Additive	Comments	Reference
Epon 812	10 ml	HHPA 11 ml	BDMA 1%	Cardolite 1 ml NC-513	DDSA replaced with reactive flexibilizer	Finck (1960)
Epon 812		DDSA + MNA	DMP-30 1.5–2.0%		Standard 1974 DDSA/MNA ratio determines hardness	Luft (1961) (Table 5.7)
Epon 812		DDSA + MNA	DMP-30 1.5%		Anhydride/epoxy ratio adjusted to improve sectioning properties	Coulter (1967) (Table 5.9)
Epon 812		DDSA + MNA	DMP-30 or BDMA		Anhydride/epoxy ratio adjusted to improve sectioning properties	Burke and Geiselman (1971)
Epon 812	10 ml	MNA 8.5 ml	DMP-30 0.15–0.3 ml	Thiokol LP-8 1.5–3.0 ml	Embedding at room temp.	Kushida (1965c)
Epon 812	8 ml	MNA 8.9 ml	DMP-30 1.5–2.0%	DER 732 2 ml	DDSA replaced with a reactive flexibilizer	Kushida (1966b)
Epon 812	2 ml	MNA 8.9 ml	DMP-30 1.5–2.0%	DER 736 8 ml	DDSA replaced with a reactive flexibilizer	Kushida (1967)
Epon 812	10 ml	NSA 13 ml	DMP-30 1.5–2.0%		Low viscosity hardener	Kushida (1971)
Epon 812 + Epon 815	10 ml	DDSA 14–16 ml	DMP-30 2 ml or BDMA		First use of Epon 812	Kushida (1959)
Epon 812 + Epon 815		DDSA	DMP-30		Epon 812/Epon 815 ratio determines hardness	Shinagawa et al. (1962) (Table 5.6)
Araldite 502 + Epon 812	20 ml 25 ml	DDSA 60 ml	DMP-30 2.5 ml	DBP 2–4 ml	First use of a mixture of Araldite and Epon	Voelz and Dworkin (1962)
Araldite CY 212 (or Araldite 502 or 6005) + Epon 812	15 ml 25 ml	DDSA 55 ml	DMP-30 1.5% or BDMA 3%		See comments in text	Mollenhauer (1964) (Table 5.8)
Araldite 506 + Epon 812	81 ml 62 ml	DDSA 100 ml	DMP-30 1.5% or BDMA 3%	DBP 4–7 ml	Popular 1974	Mollenhauer (1964)

Table 5.2 (continued)

Epoxy resin		Hardener		Accelerator		Additives		Features	Reference
Araldite 502 + Epon 812	20 ml 25 ml	DDSA	60 ml	DMP-30	1.5–2.5%			Sections susceptible to action of hydrolases	Anderson and André (1968)
Araldite 502 + Epon 812	35 ml 40 ml	DDSA	40 ml	DMP-30	1.5–2.5%				Anderson and André (1968)
Epok 533	10 ml	MNA	8.6 ml	DMP-30	1.5–2.0%	Carbowax 200	0–2%	Epoxy resin manufactured in Japan. Low viscosity	Kushida (1963b)
Maraglas 655				BDMA	2%	Cardolite NC-513 + DBP		First use of Maraglas. Hardness varied with content of flexibilizer and plasticizer	Freeman and Spurlock (1962) (Table 5.4)
Maraglas 655	68 ml			BDMA	2 ml	Cardolite NC-513 + DBP	20 ml 10 ml	Modified procedure	Spurlock et al. (1963)
Maraglas 655	36 ml			BDMA	1 ml.	DER 732 + DBP	8 ml 5 ml	Cardolite NC-513 replaced with DER 732	Erlandson (1964)
Maraglas 655	48 ml			DMP-30	2 ml	Cardolite NC-513 + TAC	40 ml 10 ml	TAC added to improve sectioning properties	Winborn (1965) (Table 5.10)
DER 332		DDSA		DMP-30		DER 732		First use of Dow epoxy resin. Hardness adjusted with content of flexibilizer	Lockwood (1964) (Table 5.5)
DER 332	7 g	DDSA	8 g	DMP-30	0.3 g	DER 732	3.2 g	Modification for amphibian eggs	Bluemink (1970)
DER 334	48 ml	DDSA	40 ml	DMP-30	2 ml	TAC	10 ml	TAC added to improve sectioning properties	Winborn (1965) (Table 5.10)
ERL 4206	10 g	NSA	26 g	DMAE	0.4 g	DER 736	6 g	Low viscosity	Spurr (1969) (Table 5.3)

Key to Table 5.2

Epoxy resins

Araldite CY 212	Manufactured by Ciba in Europe. Previously named Araldite M. Sold by Fluka (see Appendix) as Durcupan ACM. Viscosity 1300–1650 cps at 25°C.
Araldite 502,6005 and 506	Manufactured by Ciba in the USA.
Epon 812 (Epikote 812 in Europe)	A glycerol-based aliphatic epoxy resin. Manufactured by Shell. Readily takes up water (hygroscopic). Viscosity 150–210 cps at 25°C.
Epon 815	Manufactured by Shell.
Epok 533	Manufactured by Oken Shoji. Low viscosity.
Maraglas 655	Manufactured by Marblette. Viscosity 500 cps at 25°C. Variability from can to can reported.
DER 332	Manufactured by Dow. A very pure epoxide.
DER 334	Manufactured by Dow. Polymerization time varies with batch of resin. Viscosity 500–700 cps. at 25°C.
ERL 4206	Vinyl cyclohexene dioxide. Manufactured by Union Carbide. Viscosity 7.8 cps at 25°C.

Hardeners

DDSA	dodecenyl succinic anhydride. Viscosity 290 cps at 25°C. Trade name HY 964 (previously 964B) (Ciba).
HHPA	hexahydrophthalic anhydride. A white crystalline solid. Melting point 35°C. Trade name Epikure HPA (Shell)
MNA or NMA	methyl nadic anhydride or nadic methyl anhydride. Viscosity 175–275 cps at 25°C. Reacts with permanganates (Reedy 1965).
NSA	nonenyl succinic anhydride. Viscosity 117 cps at 25°C.

Accelerators

DMP-30	2,4,6-tridimethylamino methyl phenol. Trade names DMP-30 and DY 064 (previously 964C) (Ciba).
BDMA	benzyl dimethylamine. Trade name DY 062 (Ciba).
DMAE	dimethylaminoethanol. Gives longer useful pot-life than DMP-30 or BDMA. Trade name S-1 (Pennsalt).

Additives

DBP	dibutyl phthalate. Plasticizer.
TAC	triallyl cyanurate
Cardolite NC-513	A long chain mono-epoxide resin flexibilizer. One epoxy group per molecule. Compatible with propylene oxide. Viscosity 50 cps at 25°C.
Thiokol LP-8	A liquid polythiodithiol polymer of low viscosity. A reactive flexibilizer.
DER 732	A polyglycol diepoxide flexibilizer. Manufactured by Dow. Viscosity 55–100 cps at 25°C.
DER 736	Digycidyl ether of polypropylene glycol. Manufactured by Dow. Shorter chain length than DER 732. Viscosity 30–60 cps at 25°C.
Carbowax 200	polyethylene glycol 200

This is tedious and time consuming. In the majority of laboratories a standard embedding medium and procedure is adopted and then used for many years without modification for a great range of specimens.

A survey of the recent literature indicates that the Epon 812 mixture of Luft (1961) is a standard embedding medium throughout the world, while Araldite CY 212 (Glauert et al. 1956) is widely used in Europe where it is easily obtainable. The mixture of Epon 812 and Araldite CY 212 or 506 proposed by Mollenhauer (1964) is also popular, and the low-viscosity embedding medium of Spurr (1969) based on ERL 4206 is valuable for specimens that are hard to infiltrate with more viscous epoxy resins.

5.3.2 CHARACTERISTICS OF EPOXY RESIN EMBEDDING MEDIA

5.3.2.a *Viscosity*

The main differences between the various epoxy resins is in their viscosity. They can be placed in the following order, ranging from high to low viscosity:

> Araldite 502 (3000)
> Araldite CY 212 (1300–1650)
> DER 334 (500–700)
> Maraglas 655 (500)
> Epon 812 (150–210)
> ERL 4206 (7.8)

The viscosity in centipoise (cps) at 25 °C is given in brackets.

Longer infiltration times are required with the more viscous media (§ 5.3.4), and for very dense specimens it is advisable to choose a medium of low viscosity.

The viscosity of the complete embedding medium is also influenced by the viscosity of the hardener, and can be reduced by using NSA (117 cps) in preference to HHPA (which is solid at 25 °C), DDSA (290 cps) or MNA (175–275 cps) (Spurr 1969; Kushida 1971). The viscosity is also less for media containing plasticizers or flexibilizers added to modify the hardness and sectioning properties of the final block.

The viscosity of an epoxy resin embedding medium usually increases on the addition of the accelerator, and then continues to increase with time at a rate depending on the temperature as polymerization proceeds. For example, Araldite mixtures become too viscous to use after about 48 hr at room temperature.

Spurr (1969) selected the epoxy resin ERL 4206 (7.8 cps), the hardener

NSA (117 cps) and the reactive flexibilizer DER 736 (30–60 cps), all of which have a low viscosity, for his low-viscosity embedding medium. The complete medium has a viscosity of 60 cps at 25 °C. This medium must *always* be used in a fume cupboard; the volatile vapours may be harmful.

TABLE 5.3
Low-viscosity epoxy resin embedding medium (Spurr 1969)

Mixture	Standard medium	Modifications			
	A	B	C	D	E
	(ml)	(ml)	(ml)	(ml)	(ml)
ERL 4206	10.0	10.0	10.0	10.0	10.0
DER 736	6.0	4.0	8.0	6.0	6.0
NSA	26.0	26.0	26.0	26.0	26.0
DMAE (S 1)	0.4	0.4	0.4	1.0	0.2
Hardness	Firm	Hard	Soft		
Polymerization time at 70°C (hr)	8	8	8	3	16
Pot life (days)	3–4	3–4	3–4	2	7

The rate of polymerization and viscosity of the medium, and the hardness of the final block, can be modified by adjusting the relative amounts of ERL 4206, DER 736 and the accelerator dimethylaminoethanol (DMAE) (Table 5.3). Modifications favouring one feature generally involve sacrifice of other qualities of the medium. For example, a medium with a rapid rate of polymerization (mixture D) has a short pot life, while a medium with a maximum pot life and minimum viscosity (mixture E) requires a longer polymerization time.

5.3.2.b Hardness of the final block
The hardness of the final block is influenced by each of the components of the embedding medium and can be modified in a number of ways.

(i) *Plasticizers and flexibilizers.* The simplest way to reduce the hardness of the final block is to add a plasticizer, dibutyl phthalate (DBP) (Glauert et al. 1956) or a flexibilizer, Cardolite NC-513 (Finck 1960), DER 732 (Lockwood 1964), Thiokol LP-8 (Kushida 1965c) or DER 736 (Kushida 1967). The flexibilizers are usually to be preferred to the plasticizers since they react with the epoxy resin and become part of the cross-linked structure. Consequently they are less likely to be lost under electron bombardment. In some embedding media, however, dibutyl phthalate is necessary to prevent the block becoming too brittle (e.g. Erlandson 1964) and to improve the sectioning properties (Mollenhauer 1964).

TABLE 5.4
Variation of hardness with the content of plasticizer and flexibilizer
(Freeman and Spurlock 1962)

Mixture	A	B	C	D	E
	(%)	(%)	(%)	(%)	(%)
Maraglas 655	60	60	65	65	70
Cardolite NC-513	40	30	30	20	20
DBP	–	10	5	15	10
	soft ―――――――――――――――――――――→ hard				

All mixtures contain 2% BDMA as accelerator.

Mixture E is generally preferred so long as care is taken to obtain adequate infiltration of the embedding medium into the specimen (Spurlock et al. 1963).

Erlandson (1964) reported that the mixture recommended by Freeman and Spurlock (1962) (Table 5.4) yielded quite brittle blocks and suggested replacing Cardolite NC-513 with the polyglycol diepoxide flexibilizer DER 732. He recommends a mixture containing Maraglas 655 (72%), DER 732 (16%), DBP (10%) and BDMA (2%). The sectioning properties of the blocks can be modified by varying the proportions of Maraglas 655 and DER 732. The hardness of the blocks decreases with increasing concentrations of DER 732; for harder blocks as little as 10% of DER 732 may be used, while softer blocks are obtained by increasing the concentration to 30%.

Lockwood (1964) tested a large group of mixtures of the epoxy resin DER 332 in combination with the reactive flexibilizer DER 732 and recommended the three mixtures listed in Table 5.5.

TABLE 5.5
Variation of hardness with the content of
reactive flexibilizer (Lockwood 1964)

Mixture	1	2	3
	(ml)	(ml)	(ml)
DER 332	7	7	6
DDSA	5	5	10
DER 732	3	2	3
DMP–30	0.30	0.28	0.38
Hardness	soft	hard	soft

Mixture No. 1 gives soft blocks and is particularly suitable for soft tissues (kidney, liver, etc.) and for isolated cells. Mixture No. 2 gives harder blocks and is preferable for tissues rich in collagen. Mixture No. 3 gives blocks with the same hardness as mixture No. 1 but gives stronger sections due to the greater concentration of DDSA. Bluemink (1970) modified the DER 332 embedding medium proposed by Lockwood (1964) in accordance with the recommendations of Coulter (1967) concerning anhydride/epoxy ratios (§5.3.2c), and used a mixture containing DER 332 (7.0 g), DDSA (8.0 g), DER 732 (3.2 g) and DMP-30 (0.3 g) for embedding amphibian eggs.

(ii) *Mixtures of epoxy resins.* In embedding media containing two epoxy resins the hardness of the block depends on the relative amounts of the two resins (Kushida 1959; Shinagawa et al. 1962).

TABLE 5.6

Variation of hardness with the ratio of two epoxy resins
(Shinagawa et al. 1962)

Ingredients	(ml)	(ml)	(ml)	(ml)	(ml)
Epon 812	4.0	5.0	6.0	7.0	8.0
Epon 815	6.0	5.0	4.0	3.0	2.0
DDSA	16.6	16.0	15.3	14.7	14.0
DMP-30	0.53	0.52	0.51	0.49	0.48
	soft				→ hard

A ratio of Epon 812/Epon 815 of 5/5 is recommended for sectioning at room temperature. Softer blocks are suitable for sectioning in the cold, and harder blocks for sectioning in the warm.

(iii) *Hardeners.* The hardness of the block also depends on the nature of the hardener. The softest blocks result from the use of DDSA and harder blocks are obtained by replacing some or all of the DDSA with HHPA (Finck 1960) or MNA (Luft 1961; Robson 1964). Luft (1961) proposed using a mixture of Epon 812 and DDSA, and a mixture of Epon 812 and MNA in various proportions to obtain blocks of different hardness.

These mixtures are very popular since they provide an easy method for varying the hardness of the block.

The hardness also increases as the concentration of the accelerator increases, but adjustment of the accelerator is not a good method of varying the hardness since the block may become brittle and difficult to section.

TABLE 5.7
Variations of hardness with the ratio of DDSA/MNA (Luft 1961)

	(ml)	(ml)	(ml)	(ml)	(ml)
Mixture A	10	7	5	3	0
Mixture B	0	3	5	7	10
BDMA	0.15	0.15	0.15	0.15	0.15

soft ————————————————————→ hard

Mixture A: Epon 812 (62 ml) and DDSA (100 ml)
Mixture B: Epon 812 (100 ml) and MNA (89 ml)

(iv) *Epon-Araldite mixture.*   Mollenhauer (1964) recommended the three embedding media listed in Table 5.8, following a study of a range of different media consisting of various combinations of the epoxy resins Epon 812, Araldite CY 212 (M) and Araldite 506, the hardener DDSA, the flexibilizer Cardolite NC-513, the plasticizer dibutyl phthalate and the accelerators DMP-30 and BDMA.

TABLE 5.8
Composition and properties of embedding media containing Epon
and Araldite (Mollenhauer 1964)

Mixture	1	2	3
	(ml)	(ml)	(ml)
Epon 812	25	62	–
Araldite CY 212 (M)	15	–	–
Araldite 506	–	81	50
DDSA	55	100	–
Cardolite NC-513	–	–	25
Dibutyl phthalate	2–4	4–7	1–2
Relative hardness	medium	soft–medium	soft–medium
Image contrast	high	medium	low
Tissue preservation	good	excellent	excellent

These mixtures were developed primarily for plant tissues, and Mollenhauer (1964) suggested that changes in the proportions of the components of each mixture may be necessary to adapt them to a given tissue. In practice these mixtures have proved to be popular for a wide range of specimens with little modification, particularly mixtures Nos. 1 and 2.

If Araldite CY 212 is not available, Araldite 502 or Araldite 6005 can be substituted for it in the same proportions with equal success. Tissue embedded in mixture No. 1 is easier to section than when embedded in mixtures containing only Epon 812, Araldite CY 212, 502 or 6005. Tissue preservation is comparable to that obtained with Epon 812 alone. Mixture No. 2 is easier to section than No. 1 and gives better tissue preservation, but shows slightly less image contrast. Mixture No. 3 is slightly more difficult to section than either No. 1 or No. 2, and image contrast is lower. Preservation of tissue is excellent, and the mixture is particularly useful for specimens, such as pollen grains, which tend to be pulled out of the block during sectioning. This type of damage is rarely seen with mixture No. 3.

The plasticizer, dibutyl phthalate, improves the cutting properties of these mixtures, but can be eliminated if a relatively hard block is desired. Either DMP-30 or BDMA can be used as the accelerator for mixtures Nos. 1 and 2. BDMA must be used for mixture No. 3, since DMP-30 causes a precipitate to form slowly in the first epoxy resin/propylene oxide mixture during infiltration (see §5.3.4). The amount of accelerator required is about 1.5% of (fresh) DMP-30 and 3% of BDMA.

5.3.2c *Sectioning properties*

(i) *Anhydride/epoxy ratio.* The sectioning properties of the final block depend not only on its hardness but also on the extent and nature of the cross-links formed during polymerization. It is generally agreed that a linear, shorter polymer with few cross-links is easiest to section (Luft 1961; Coulter 1967), although there is some disagreement about the means of achieving this. Luft (1961) suggested that adjustment of the curing schedule will influence the sectioning properties, but Coulter (1967) found that this had no effect, and that the ratio of anhydride to epoxy was the important factor. The sectioning properties improved as the ratio was lowered from 1.0 to 0.6 and Coulter concluded that a reduction in anhydride favours a shorter polymer with fewer cross-links. The recommended modifications to Luft's (1961) Epon 812 embedding medium (Table 5.7) are as follows:

TABLE 5.9

Adjustment of the anhydride/epoxy ratio to
improve sectioning properties (Coulter 1967)

Mixture A	Epon 812 (100 ml) and DDSA (124 ml)
Mixture B	Epon 812 (100 ml) and MNA (67 ml)

The embedding medium consists of A and B mixed in the proportion of 1.5 to 1.0 with the addition of 1.5% DMP-30.

Using the same principles Coulter (1967) also calculated that an embedding medium based on Araldite 502 should contain 7.8 ml of DDSA to each 10 ml of Araldite 502 with 1.5% of DMP-30.

The calculations made by Coulter (1967) assumed that the weight per epoxide (WPE) (that is the number of grams of resin which contain an average of 1.0 g equivalent of epoxide) is 159 for Epon 812, and 502 for Araldite 502. Burke and Geiselman (1971) have found, however, that commercial samples of Epon 812 have considerable variations in WPE (from 140 to 160), and suggest that the correct proportions of hardener to epoxy resin should be calculated for each batch of resin. To be able to do this it is necessary to know the WPE; this is printed on the bottle by some suppliers (e.g. Ladd, see Appendix), and will be provided by other suppliers on request. Burke and Geiselman (1971) describe a method of calculating the correct proportions of Epon 812, DDSA and MNA to use with any particular batch of resin, and provide a table listing these proportions for any particular WPE. A typical mixture contains 27.1 g of Epon 812, 7.6 g of DDSA and 15.3 g of MNA, with the addition of 0.5 to 1.5% of BDMA or DMP-30. To take full advantage of these calculations it is essential to measure the weights of the various components very accurately. This is very difficult when the components are viscous and it is doubtful whether it is necessary for such measures to be taken in the majority of routine studies. It is suggested that electron microscopists only resort to such calculations when serious sectioning difficulties are encountered.

(ii) *Softening temperature*

TABLE 5.10

Addition of TAC to improve sectioning properties (Winborn 1965)

Mixture	1	2	3
Ingredient by volume	(%)	(%)	(%)
DER 334	48	–	–
Maraglas 655	–	–	48
Araldite 502	–	68	–
DDSA	40	19	-
Cardolite NC-513	–	–	40
TAC	10	10	10
DMP-30	2	3	2
Relative hardness	soft-medium	soft-medium	medium
Image contrast	high	high	high
Thermal stability	excellent	good	excellent

Sectioning problems can also arise if the block softens at normal laboratory temperatures, and Winborn (1965) suggest that triallyl cyanurate (TAC) should be added to epoxy resin embedding media to give polymers with a high softening temperature. The nature of the reaction between TAC and the other components of the embedding medium is not known.

The hardness of the block decreases as the concentration of TAC increases. The mixture containing Maraglas 655 (No. 3) yields blocks which are slightly harder and more difficult to section than blocks containing DER 334 or Araldite 502.

5.3.3 PREPARATION OF EPOXY RESIN EMBEDDING MEDIA

Epoxy resins and their hardeners are supplied in cans or bottles and can be stored indefinitely at room temperature so long as they are kept *dry*. Storage in glass-stoppered bottles is usually preferred. The epoxy resin ERL 4206 and the epoxy flexibilizer DER 736 will dissolve the plastic liners of some screw caps and should be stored in bottles with glass or cork stoppers (Spurr 1969). Some samples of the accelerators are less stable and may be inactivated if they are allowed to take up water. They should be stored in bottles with well-fitting caps in a desiccator at room temperature.

Note: Epoxy resins, hardeners and, in particular, accelerators are toxic and may cause skin irritation and dermatitis. They should be handled with great care in an area with good ventilation and preferably in a fume cupboard. Repeated contact with the skin should be avoided. The polymerized resin is inert.

Most of the epoxy resin embedding media take a long time to harden at room temperature, that is they have a long 'pot life', but they gradually become more viscous. Consequently, it is advisable to prepare the embedding medium just before use when the specimens are in the final stages of dehydration.

It is essential to mix the components of the embedding medium very thoroughly to obtain uniform polymerization. Inadequate mixing is one of the main causes of sectioning problems with epoxy resins. The simplest way to obtain good mixing is to warm the epoxy resin, the hardener *and* the containers to be used for measuring and mixing to 60 °C (Glauert and Glauert 1958). The hardener HHPA is solid at room temperature and must warmed to 55–60 °C before mixing with the epoxy resin. The warm components are easily poured from the containers in which they have been stored into a (warm) graduated cylinder, for measurement by volume, or into a (warm) beaker, on the pan of a balance, for measurement by weight. The mixture is immediately poured into a (warm) conical flask containing the required amount of accelerator. In general slight variations in the relative amounts of the various components are not important, except for the accelerator which should be measured accurately. A simple method is to calibrate a Pasteur pipette in 'drops of accelerator per ml' and to keep it for adding the accelerator. Alternatively a plastic syringe can be used. BDMA is much less viscous than DMP-30 and is much easier to handle. Care must be taken not to add too much accelerator or the block may be brittle and difficult to section.

The warm embedding medium in the conical flask is mixed by swirling and/or stirring with a warm glass rod. A uniform mixture is obtained in a few minutes by this method and no special stirring apparatus is required. A few bubbles may form during stirring but these will go if the mixture is allowed to stand at 60 °C for a short time. If the components of the embedding medium are not warmed then a much longer mixing time is required and it is advisable to use one of the stirrers specially designed for the purpose (e.g. Zacks 1963).

It is very difficult and tedious to wash glassware after it has been in contact with epoxy resins and it is convenient to use disposable polyethylene containers whenever possible since the resins do not stick to polyethylene. Plastic syringes are useful when it is necessary to measure small volumes of the components accurately. If glassware is used it is advisable to keep a separate set of glassware for the purpose, and to only attempt to clean the container used for measuring the amounts of the components (measuring cylinder or beaker). The mixing procedure is unaffected by a thin film of hardened resin on the surfaces of the other containers. Cleaning should be done in acetone or ethanol as soon as possible after use.

It is best to prepare the complete mixture just before use but if necessary it can be stored at 4 °C in a bottle with a well-fitting cap for several weeks, or for many months at − 20 °C. During infiltration of the specimens the embedding medium is kept at room temperature and warmed to 60 °C and stirred well before each change of solution.

5.3.4 EMBEDDING SCHEDULES FOR EPOXY RESINS

At the end of dehydration the specimens will be immersed in the intermediate solvent (usually propylene oxide) (see §4.2) in small vials (Fig. 3.2).

5.3.4.a *Standard embedding schedule*
1. Remove the solvent with a pipette and flush it down a sink in a fume

cupboard with a large volume of water. If propylene oxide is being used do not remove it all since it is very volatile and the specimen may dry out.

2. Replace the solvent with a 1/1 mixture of solvent and embedding medium. Shake the vial gently to mix the components and leave for 30 min to 1 hr at room temperature.

3. Remove the mixture with a pipette and replace it with embedding medium. Leave 16 to 24 hr at room temperature with the cap off the vial to allow any remaining solvent to evaporate. Infiltration of dense specimens is aided by placing the vials on a shaker (Figs. 5.1 and 5.2).

4. Transfer each specimen to a *dry* capsule and fill the capsules with embedding medium. Insert labels (Fig. 5.5).

5. Place the capsules in an oven overnight at 60 °C (§ 5.2).

This schedule is given as a guide line for embedding. It is widely used for embedding in Epon 812 or Araldite CY 212, but it is clear from the published work (see references in Table 5.2) that the embedding schedule is not critical and that considerable variations can be introduced with equally successful results for the great majority of specimens. The original schedules recommended for the various epoxy resin embedding media are listed in Table 5.11.

Epoxy resins should not be poured down a sink after use since they may cause a blockage. Used embedding medium should be drained into a small box (an old plate box is suitable) and allowed to harden before being discarded.

Most epoxy resin embedding media harden overnight at 60 °C and this is a convenient temperature to use. Approximately 2 days are usually required for hardening at 50 °C, while the time can be reduced to 8 hr or less by raising the temperature to 70 °C or higher (e.g. Coulter 1967; Spurr 1969). The sectioning properties of most blocks improves with time and the capsules can safely be left in the oven for some days after hardening.

Luft (1961) suggested that a series of increasing temperatures (overnight at 35 °C, a day at 45 °C, and then overnight at 60 °C) be used for polymerizing epoxy resins, but there is little evidence that the resultant blocks are any different from those placed directly at 60 °C.

Ultra-violet (UV) irradiation can be used as an alternative to heat for the polymerization of epoxy resins (Shinagawa et al. 1962) with the advantage that polymerization can be carried out at a lower temperature if necessary (§ 5.2). Shinagawa et al. (1962) compared the hardening times of a mixture of Epon 812 and Epon 815 with heat and with UV and found that the times

TABLE 5.11

Embedding schedules for epoxy resins

	p.o./Epoxy (2/1)	(1/1)	(1/2)	Embedding medium	Polymerization		Reference
Standard schedule	–	30–60 min	–	16–24 hr	overnight	60°C	Parker (1972)
Epok 533	–	60 min	–	1–2 hr; 1–2 hr	50 hr	50°C	Kushida (1963b)
Maraglas 655	–	30 min	–	12 hr (at 10°C)	24–48 hr	60°C	Spurlock et al. (1963)
Maraglas 655		45 min	–	1 hr; 2–3 hr	17 hr	52°C	Erlandson (1964)
DER 332		1 hr (at 45°C)	–	1 hr (at 37°C)	3 × 24 hr	37°C, 45°C, 60°C	Lockwood (1964)
Epon 812 + Araldite	1 hr	–	1 hr	2–4 hr	overnight	40°C–80°C	Mollenhauer (1964)
Media + TAC Araldite 502 Maraglas 655 DER 334	–	30 min	(1/3) 30 min	1 hr	4 hr + 24 hr + 8 hr + 20–40 hr	45°C 65–70°C 65–70°C 65–70°C	Winborn (1965)
ERL 4206	–	30 min	(1/3) 30 min	3–4 hr; overnight (large specimens only)	8–16 hr	70°C	Spurr (1969)

Key: p.o., propylene oxide; TAC, triallyl cyanurate

were 60 hr and 48 hr respectively at 50 °C, 170 hr and 70 hr at 40 °C, and infinity and 100 hr at 30 °C.

The standard embedding schedule is adequate for the majority of specimens, but sometimes the specimen is found to be softer than the surrounding embedding medium as a result of incomplete infiltration. This is more likely to happen with large, dense specimens and when thick cell walls are present. Adequate infiltration can be achieved using some or all of the following modifications:

(a) Make sure that the dehydration is complete by increasing the time of soaking in the intermediate solvent.

(b) Use an embedding medium of low viscosity (e.g. Spurr 1969).

(c) Pass the specimens more gradually from the intermediate solvent to the embedding medium using ratios of 2/1, 1/1, 1/2, etc.

(d) Allow a longer infiltration time at stage 3 of the standard schedule and place the vials on a shaker. Change the embedding medium at the beginning and end of each day.

(e) De-gas the specimens in a vacuum desiccator for at least 30 min before polymerization (Erlandson 1964) or reduce the pressure in the oven to 0.5 atmospheres for a part or all of the incubation period (Winborn 1965).

The final blocks can be stored indefinitely at room temperature and their sectioning properties continue to steadily improve. Bubbles tend to form between the gelatin capsule and hardened Maraglas when the capsules are left at room temperature for a long time. Spurlock et al. (1963) advise removing the blocks from the gelatin capsules soon after removal from the oven. Bubble formation does not occur with polyethylene capsules or with other epoxy resins, so long as the capsules are thoroughly dried before use.

5.3.4.b *Embedding at room temperature*

The high temperatures used in the standard embedding schedule (§5.3.4a) are unsatisfactory for specimens for some histochemical studies. Kushida (1965c) developed a method of embedding at room temperature in a medium consisting of Epon 812 (100 ml), MNA (85 ml), Thiokol LP-8 (15–30 ml) and DMP-30 (3 ml) (Table 5.2). This mixture hardens in about 7 days at 20 °C, and in about 4 days at 30 °C. The temperature of the specimen is about 0.5 °C higher than that of the oven at 20 °C, and about 0.8 °C higher at 30 °C during the early stages of polymerization. Other epoxy resin embedding media can be polymerized at room temperature (approx. 22 °C) using ultra-violet irradiation (§5.3.4a).

Schedules for embedding at very low temperatures are described in (§6.1).

5.3.4c   Rapid embedding

The embedding procedure can be speeded up so long as the specimens are small (less than 0.1 mm in one dimension). Coulter (1967) used the rapid dehydration schedule described in §4.2.2 and then embedded specimens rapidly in Epon 812 or Araldite 502. The last change of propylene oxide was poured off, a 1:2 mixture of propylene oxide and epoxy resin added, and the specimens agitated for 5 min. The specimens were then transferred to the centres of capsules containing embedding medium which had been previously evacuated. The infiltration was completed in 5 min at the maximum vacuum available with a water aspirator. The embedding medium was polymerized at 95°C for 40 min to 12 hr (for Araldite) or for 3 hr to 12 hr (for Epon).

Robbins and Jentzsch (1967) infiltrated monolayers of cells rapidly with Epon 812, using three 5 min changes of the complete embedding medium, following rapid dehydration in ethanol and Epon 812 (§4.2.3). Polymerization was not carried out at a high temperature and took 3 days at 55°C.

Hayat and Giaquinta (1970) also developed a rapid embedding method with Epon 812 and found that adequate infiltration was obtained with two changes of Epon of 10 min each without the application of a vacuum. The embedding medium hardened in 1 hr at 100°C and the blocks could be sectioned 15 min after removal from the oven.

5.4   Polyester resins

Polyester resins were introduced as embedding media for electron microscopy by Kellenberger et al. (1956) and Ryter and Kellenberger (1958a, b) in Switzerland and by Kushida (1960a) in Japan. They have similar excellent characteristics as embedding media to the epoxy resins but are less widely used because they are not so readily available and some of the components of the embedding medium are unstable and have to be replaced every few months.

5.4.1   POLYESTER RESIN EMBEDDING MEDIA

The earliest polyester embedding medium based on Vinox K3 (Kellenberger et al. 1956) was soon superceded by Vestopal W (Ryter and Kellenberger 1958a, b) which has remained the standard polyester embedding medium outside Japan. The embedding medium contains tertiary butyl perbenzoate or benzoyl peroxide (BP) as an initiator of polymerization and cobalt naphthenate (CN) as an accelerator (Table 5.12). BP and CN may explode if mixed together and this leads to problems in the preparation of the

Polyester resin embedding media

Polyester resin	Initiator or catalyst		Activator or accelerator		Comments	Reference
Vestopal W	TBP or BP	1%	CN	0.5%	First polyester resin embedding medium	Ryter and Kellenberger (1958a, b)
Vestopal W	TBP	1%	CN	0.5%	10% TAC added to improve sectioning	Winborn (1963)
Vestopal W	BPP	1%	—		Activator omitted	Kushida (1964b)
Vestopal W	benzoin	0.3%	—		For polymerization by UV	Kushida (1964b)
Rigolac 2004 + Rigalac 70F	BPP	1%	—		Low viscosity. Ratio of 2004/70F determines hardness	Kushida (1960a)
Rigolac 2004 + Rigolac 70F	benzoin	0.5%	—		For polymerization by UV	Kushida (1961c)
Styrene/Rigolac 70F (7/3) or Styrene/Rigolac 2004 (6/4)	BP	1%	—		Ratio of styrene/Rigolac determines hardness	Shinagawa and Uchida (1961)
Rhodester 1108	Butanox	1%	—		Rhodester; available in Paris; contains styrene	Argagnon and Enjalbert (1964)
Beetle 4116 22.5 ml + Beetle 4134 7.5 ml	Butanox M 50	0.3 ml	NL 49/ST or Q2	0.3 ml 0.3 ml	For rapid embedding	Rampley and Morris (1972)

Key:
TBP   tertiary butyl perbenzoate
BP   benzoyl peroxide
BPP   benzoyl peroxide paste containing 50% benzoyl peroxide in tricresyl phosphate (Kushida 1960a). Manufactured by Riken Goseijushi in Japan. Trade name Luperco ATC.
benzoin   phenyl-benzoyl carbinol
Butanox M 50 A mixture of methyl ethyl ketone peroxides in a phthalate plasticizer
CN   cobalt naphthenate
NL 49/ST   a styrene-based solution containing 1% cobalt
Q2   a vanadium-based solution
TAC   triallyl cyanurate
UV   ultra-violet irradiation

embedding medium (§5.4.2). Kushida (1964b) has shown that Vestopal W can be polymerized by ultra-violet irradiation with the addition of benzoin as a catalyst, with the advantage that polymerization can be carried out at room temperature. Kushida (1964b) also obtained polymerization by heat with benzoyl peroxide in the absence of cobalt naphthenate. Winborn (1963) recommended the addition of 10% triallyl cyanurate to the Vestopal W embedding medium of Ryter and Kellenberger (1958a) to improve the sectioning properties of the final block.

The polyester resins Rigolac 2004 and Rigolas 70F manufactured in Japan (Kushida 1960a) also polymerize when heated in the presence of benzoyl peroxide alone, and can also be polymerized by UV irradiation (Kushida 1961c). The Rigolac mixture has the advantage that it is less viscous than Vestopal W and the hardness of the final block can be adjusted by varying the ratio of Rigolac 2004:Rigolac 70F from 8:2 to 6:4. The block has a yellow fluorescence which increases with the degree of polymerization and this is a convenient means of ascertaining how the polymerization is progressing (Kushida 1961c).

The other polyester resin embedding media described in the literature are Selectron (Low and Clevenger 1962) which was used in a mixture with *n*-butyl methacrylate (see §5.5), Rhodester (Argagnon and Enjalbert 1964), which is based on components locally available in France, and Beetle (Rampley and Morris 1972), which was recommended for rapid embedding since it hardens much faster than Vestopal W. The Beetle embedding medium was only introduced recently and it will be interesting to see if it becomes widely used.

A mixture of styrene and Rigolac (Table 5.12) can also be used as an embedding medium (Shinagawa and Uchida 1961). The hardness of the final block can be varied by altering the styrene/Rigolac ratio.

5.4.2 PREPARATION OF POLYESTER RESIN EMBEDDING MEDIA

Polyester resins, unlike epoxy resins, are polymerized by light, heat and oxygen (Kushida 1960b). They should be stored in bottles with tightly-fitting caps in a refrigerator and well protected from the light. They are then stable for many months. Benzoyl peroxide, cobalt naphthenate and the other catalysts and accelerators are easily inactivated and should be kept dry and away from the light in a refrigerator. Many workers recommend replacing stocks of the initiator and the accelerator every two months.

In general it is preferable to prepare the embedding medium just before use, although Kurtz (1961) reported that the complete Vestopal embedding mixture can be stored for a week in the refrigerator and that it will not polymerize at room temperature for 4 days.

As with epoxy resins (§ 5.3.3) thorough mixing of the viscous embedding medium is essential, and the use of a stirrer or mixer is advised (§ 5.2). When preparing an embedding medium containing cobalt naphthenate, the initiator (benzoyl peroxide) should be mixed well with the Vestopal W for 20 to 30 min before the addition of the cobalt naphthenate. The mixture is then stirred for a further 20 to 30 min (Estes and Apicella 1969). These precautions are necessary because benzoyl peroxide and cobalt naphthenate may explode if mixed directly together. Similarly the catalyst and accelerator in the Beetle embedding medium (Table 5.12) must not be mixed together before being added to the polyester resin (Rampley and Morris 1972).

5.4.3 EMBEDDING SCHEDULES FOR POLYESTER RESINS

Polyester resins are not soluble in ethanol and it is customary to dehydrate specimens in carefully dried acetone. Alternatively the specimens are dehydrated in ethanol and then acetone or styrene (Kurtz 1961; Shinagawa and Uchida 1961) is used as an intermediate solvent. Two changes of solvent of 30 min each are usual. Methyl methacrylate can also be used as an intermediate solvent (Kushida and Fujita 1968) with the advantage that it reacts with Vestopal W plus benzoyl peroxide and becomes an integral part of the polymer.

Polyester resins are viscous and are handled in the same way and with the same precautions as epoxy resins (§ 5.3.4). A shaker or agitator should be used to aid infiltration (§ 5.2) which is done at room temperature.

5.4.3.a *Embedding schedule for Vestopal W*

(Ryter and Kellenberger 1958b; Kushida 1964b)

acetone/Vestopal W	3/1	30–60 min
acetone/Vestopal W	1/1	30–60 min
acetone/Vestopal W	1/3	30–60 min
Vestopal W + catalyst		several hours
polymerization		12–24 hr at 60 °C (BP + CN as catalysts)
	or	10 hr at room temperature under UV (benzoin as catalyst).

For large or dense specimens better infiltration is obtained by a more gradual replacement of the acetone with Vestopal. The specimens are placed in a vial with a few drops of acetone and then the Vestopal is added drop by drop (De Haller et al. 1961). One drop of Vestopal (plus catalyst) is added every 30 min and is mixed well by shaking the vial. After 10 to 12 drops have been added, the Vestopal is replaced with fresh Vestopal and this is left overnight at room temperature. The next day the Vestopal is again replaced with fresh Vestopal and this is left for several hours before the specimens are embedded in capsules in the usual way.

Vestopal is more readily soluble in styrene than in acetone and when styrene is the intermediate solvent a soak in a 1/1 mixture of styrene/Vestopal W for 30 min under a moderate vacuum, followed by several hours (or overnight) in Vestopal W plus catalyst, is sufficient to obtain adequate infiltration (Kurtz 1961).

5.4.3.b   *Rapid embedding with Vestopal W*
           (Estes and Apicella 1969)

Following dehydration in ethanol and two rinses in styrene of 5 min each:

Styrene/Vestopal W	1/1	10 min
Vestopal W + BP + CN		10 min
Vestopal W + BP + CN		10 min
Polymerization		1 hr at 45–50 °C in a vacuum oven (15″ Hg)
	then	15 min at 70–80 °C
	then	15–20 min at 90–100 °C

The blocks are ready for sectioning as soon as they have cooled to room temperature.

The Beetle embedding medium developed by Rampley and Morris (1972) is suitable for rapid embedding since the mixture polymerizes in 1 hr at 60 °C.

5.4.3.c   *Embedding schedule for Rigolac*
           (Kushida 1960a, 1961c)

acetone/Rigolac + catalyst	1/1	1–2 hr
Rigolac + catalyst		1–2 hr
Rigolac + catalyst		1–2 hr
polymerization		18–24 hr at 55 °C (BP as catalyst)
	or	4–10 hr under UV (benzoin as catalyst)

5.4.3.d   *Embedding schedule for styrene/Rigolac*
(Shinagawa and Uchida 1961)

acetone/styrene	1/1	30 min
styrene		30 min
styrene/Rigolac + BP 7/3		1–2 hr
or 6/4		
polymerization		overnight at 50–60 °C

5.5   Methacrylates

The methacrylates were widely used as embedding media in electron micro-scopy from 1949 when they were introduced by Newman et al. until the development of media based on epoxy and polyester resins in the late 1950s (§5.3 and §5.4). The methacrylates have major disadvantages as embedding media. In particular they polymerize with considerable shrinkage (up to 20%) and are unstable in the electron beam (Reid 1974). Consequently they are no longer used as routine embedding media, although the water-soluble metha-crylates (§5.6.2) have proved to be valuable for cytochemical studies. Only a brief description of the methods of embedding in methacrylates is included here. The reader is referred to earlier textbooks (Pease 1964; Glauert 1965) for fuller details.

5.5.1   METHACRYLATE EMBEDDING MEDIA

Methacrylate embedding media consist of a mixture of *n*-butyl and methyl methacrylates with benzoyl peroxide, Luperco or 2,2-azo-bis-iso-butyronitrile (AIB) as a catalyst for polymerization by heat, or with AIB, benzoin or uranyl nitrate (Kushida 1962c) for polymerization by ultra-violet irradiation (see review by Kushida 1965b). The hardness of the final block is adjusted by varying the proportion of *n*-butyl to methyl methacrylate, the hardness increasing with the proportion of methyl methacrylate.

Impurities in the embedding medium, or even the specimen itself, can initiate polymerization of methacrylates, so that polymerization may start in one region before others. The addition of a small quantity of an initiator, uranyl nitrate (0.01%) (Ward 1958), to the embedding medium aids more uniform polymerization.

The large shrinkage that occurs during polymerization of methacrylates may damage the specimen, and it is therefore customary to use partially polymerized methacrylate for the final infiltration and embedding. Part of the shrinkage will then have occurred before the specimen is placed in the embedding medium. Unfortunately, partially polymerized methacrylate is very viscous and the advantage of easy infiltration by low viscosity methacrylate is lost.

The stability of methacrylate sections under electron bombardment is improved by the addition of the cross-linking agent divinyl benzene to the embedding medium. The polymerized methacrylate has a three-dimensional structure similar to epoxy and polyester resins (Kushida 1961b). The shrinkage during polymerization is not reduced by the addition of divinyl benzene (Kushida 1962b), and the blocks tend to split vertically during polymerization (Watson and Aldridge 1961). Fortunately this split never passes through the specimen.

Greater stability in the electron beam is also obtained by using a mixture of *n*-butyl methacrylate and styrene as the embedding medium (Kushida 1961a). Copolymerization of methacrylate and styrene probably occurs, leading to a cross-linked structure similar to that obtained with divinyl benzene (Mohr and Cocking 1968). Methacrylate-styrene mixtures also shrink during polymerization. The hardness of the final block can be adjusted by varying the ratio of *n*-butyl methacrylate to styrene from 7:3 to 4:6 (Kushida 1961a). DeLamater et al. (1971) describe a method of embedding in styrene with the addition of only 5–10% of *n*-butyl methacrylate.

Methacrylates can also be copolymerized with polyester resins (Low and Clevenger 1962).

5.5.2 PREPARATION OF METHACRYLATE EMBEDDING MEDIA

The methacrylates are clear liquids of low viscosity. They are volatile and have an unpleasant smell; they should always be handled in a fume cupboard and great care should be taken not to inhale the fumes which can be harmful. They are supplied with an inhibitor, hydroquinone, to prevent polymerization during transport and storage. This inhibitor is removed by shaking the methacrylate with sodium hydroxide in a separating funnel (Pease 1964; Glauert 1965), followed by distillation under vacuum, if divinyl benzene or styrene is to be added to the embedding medium. Methyl methacrylate is distilled at 45 °C and 100 mmHg pressure, and *n*-butyl methacrylate at 52 °C and 11 mm Hg pressure (Kushida 1961b).

It is essential that the methacrylate be kept completely dry; it should be stored in bottles with tight-fitting lids and containing a drying agent (silica gel, molecular sieve or anhydrous calcium chloride) at room temperature in a fume cupboard. Divinyl benzene and styrene are polymerized by light, heat and oxygen (Kushida 1961a, b) and should be stored in a refrigerator well protected from the light. Any space in the containers should be filled with nitrogen. Styrene is supplied with an inhibitor, but this is not removed. More reproducible results are obtained with the inhibitor still present (Mohr and Cocking 1968).

Methacrylate embedding media mix easily because of their low viscosity. For a standard medium the *n*-butyl and methyl methacrylates are mixed in the proportions required (e.g. 8:2) to give the desired hardness to the final block, and then the catalyst (e.g. 1–2% benzoyl peroxide) is added and mixed well. Most catalysts contain water and it is important to dry the embedding medium before use. The complete embedding medium plus catalyst can be stored for some weeks at 4 °C.

Partially polymerized methacrylate is prepared by heating the methacrylate plus catalyst in a conical flask at 90 °C in a water bath until the mixture thickens. The mixture is agitated by continuously shaking the flask and the flask is immediately transferred to cold water when the mixture has the required viscosity.

Embedding media containing divinyl benzene are prepared with *n*-butyl and methyl methacrylate (100 ml), divinyl benzene (5 ml) and benzoyl peroxide (1 g) (Kushida 1961b), or with *n*-butyl methacrylate (distilled) (100 ml), divinyl benzene solution (3 ml), α-terpineol (3 ml) and benzoyl peroxide (1 g) (Watson and Aldridge 1961).

Methacrylate-styrene embedding media are prepared by mixing distilled *n*-butyl methacrylate and styrene in the required ratio (e.g. 7:3) and then adding 1% benzoyl peroxide as catalyst. The embedding medium is dried and filtered before use.

5.5.3 EMBEDDING SCHEDULES FOR METHACRYLATES

Methacrylates are readily soluble in ethanol and acetone and no intermediate solvent is required. The standard embedding schedule is:

ethanol/methacrylate + catalyst	1 hr
methacrylate + catalyst	1 hr
methacrylate + catalyst	1 hr
polymerization	overnight at 60 °C
or	24–48 hr under UV

Methacrylates are volatile and so the capsules are filled completely and the caps placed on.

As with other embedding media, polymerization by ultra-violet (UV) irradiation has the advantage that it can be carried out at room temperature or below (Shinagawa and Uchida 1961).

A similar infiltration schedule is used for embedding media containing divinyl benzene or styrene. With divinyl benzene polymerization is carried out for 24 hr at 50 °C (Kushida 1961b) or for 12 hr under UV at room temperature (Watson and Aldridge 1961). Methacrylate–styrene embedding media require 30 hr under UV irradiation (Kushida 1962a) or 18–24 hr at 55 °C for polymerization if the inhibitor has been removed from the styrene (Kushida 1961a) or 44 hr at 55 °C if the inhibitor is still present (Mohr and Cocking 1968). The slowing down of the polymerization by the inhibitor is an advantage as it permits thorough infiltration before hardening begins.

A longer infiltration procedure lasting 24 hr or more is required when using viscous, partially polymerized methacrylate. A higher temperature of polymerization (70–80 °C) is recommended to reduce the viscosity of the embedding medium and aid infiltration.

5.6   Water-soluble embedding media

Embedding media that are soluble in water can be used as dehydrating agents thus avoiding the use of ethanol, acetone and propylene oxide and so preserving some components of the specimen which would otherwise be extracted. However, these embedding media also cause extraction, although the components extracted may be different. In addition, the embedding media may react with the specimen and modify it. For example, epoxy compounds in aqueous solution react readily with proteins and nucleic acids and therefore tend to act as fixatives (Gibbons 1959). This makes water-soluble epoxy resins unsuitable for experiments on the digestion of thin sections by specific enzymes, and the water-soluble methacrylates are usually preferred (Bernhard 1966). However, these methacrylates are powerful lipid solvents for neutral lipids (Cope and Williams 1968) and for some phospholipids (Cope and Williams 1969b) even at low temperatures (-20 °C).

5.6.1   WATER-SOLUBLE EPOXY RESINS

The first water-soluble epoxy resin embedding medium to be developed was Aquon (Gibbons 1959), the water miscible fraction of Epon 812. Aquon is not available commercially and has to be extracted from Epon 812 in the laboratory by a rather lengthy procedure (see Glauert 1965 for the method of extraction). Consequently, it has been largely superceded by the Ciba epoxy resin Durcupan marketed by Fluka (see Appendix). The original

Water-soluble epoxy resin embedding media

Epoxy resin	Hardener	Accelerator	Additives	Comments	Reference
Aquon 10 ml	DDSA 25 ml	BDMA 0.35 ml	–		Gibbons (1959)
Aquon 10 ml	DDSA + HHPA 19 ml 3 ml	BDMA 0.30 ml	–	For harder specimens	Gibbons (1959)
Durcupan 5 ml	DDSA 11.7 ml	DMP-30 1.0–1.2 ml	DP 0.2–0.4 ml	DP not essential	Stäubli (1960); Leduc and Bernhard (1961)
Durcupan	MNA	DMP-30 1.5–2.0%	Cardolite NC-513	Hardness varied with amounts of MNA and Cardolite	Kushida (1964a)
Durcupan	DDSA + MNA	DMP-30 1.5–2.0%		Hardness varied with ratio of DDSA/MNA	Kushida (1964a)
Durcupan 100 ml	MNA 120 ml	DMP-30 1.5–2.0%	Thiokol LP-8 20–35 ml	Hardness varied with Thiokol	Kushida (1966a)
Epon 812 20 ml	HHPA 16 ml	BDMA 1.5–2.0%			Craig et al. (1962)

Key:

Aquon   A water-miscible fraction extracted from Epon 812.
Durcupan   An aliphatic polyepoxide. A water-soluble Ciba epoxy resin originally named X133/2097.
DDSA   dodecenyl succinic anhydride.
MNA   methyl nadic anhydride
HHPA   hexahydrophthalic anhydride
BDMA   benzyl dimethyl amine
DMP-30   2,4,6-tridimethylaminomethyl phenol
DP   dibutyl phthalate

For further details see the key to Table 5.2.

Durcupan embedding medium proposed by Stäubli (1960) (Table 5.13) is very difficult to section, and has been modified by Kushida (1964a, 1966a) to produce blocks with better sectioning properties. Kushida proposed the use of MNA as hardener with Cardolite NC-513 or Thiokol LP-8 as flexibilizer, or the use of a mixture of DDSA and MNA as hardener (Table 5.13). The hardness of the final block can be adjusted by altering the amounts of the various components of the embedding media.

A dehydration and embedding schedule based on Epon 812 (and not the water-soluble component Aquon) has been proposed by Craig et al. 1962.

The water-soluble epoxy resins are handled in exactly the same way and with the same precautions as other epoxy resins (§5.3.3 and §5.3.4).

5.6.1a   *Dehydration and embedding schedule for Durcupan*
(Kushida 1964a)

A typical schedule is as follows:

50% Durcupan in water		15–30 min
70% Durcupan in water		15–30 min
90% Durcupan in water		15–30 min
Durcupan		30–60 min
Durcupan		30–60 min
Durcupan/Durcupan + hardener + accelerator	3/1	1 hr
Durcupan/Durcupan + hardener + accelerator	1/1	1 hr
Durcupan/Durcupan + hardener + accelerator	1/3	1 hr
Durcupan + hardener + accelerator		1–2 hr
Durcupan + hardener + accelerator		1–2 hr
Polymerization		50 hr at 50 °C

The specimens are dehydrated and infiltrated with Durcupan at room temperature with constant agitation on a mechanical shaker (§5.2). 30% Durcupan in water should be avoided as the tissue tends to swell (Leduc and Bernhard 1961).

5.6.1b   *Dehydration in Durcupan followed by conventional embedding*
An alternative solution to the difficulty of sectioning the original Durcupan

embedding medium (Stäubli 1960) is to embed the specimens in a conventional embedding medium after dehydration in Durcupan. Embedding is still achieved without passing through highly reactive solvents such as propylene oxide. Specimens dehydrated in Durcupan have been successfully embedded in epoxy resins (Kushida 1963a; Stäubli 1963; Bird 1964) and polyester resins, methacrylate and styrene (Kushida 1965a). The specimens are passed from Durcupan to the embedding medium through a series of mixtures of increasing concentration of the embedding medium (e.g. 3/1, 1/1, 1/3 for 1 hr each). It is important to remove all the Durcupan when embedding in polyester resins, methacrylate or styrene, because it does not react with these embedding media and will soften the final block. Durcupan copolymerizes with other epoxy resins becoming part of the polymer. Following infiltration with the embedding medium polymerization is carried out in the usual way.

5.6.2 WATER-SOLUBLE METHACRYLATES

The first water-soluble methacrylate to be used as a combined dehydrating agent and embedding medium for electron microscopy was based on glycol methacrylate (GMA) (2-hydroxyethyl methacrylate) (Rosenberg et al. 1960; Leduc et al. 1963). In the period up to 1965 the quality of commercially available GMA varied considerably, and 2-hydroxypropyl methacrylate (HPMA) was investigated as an alternative (Leduc and Holt 1965). GMA of high quality is now available and is preferred to HPMA for most purposes because HPMA penetrates more slowly and the dehydration and embedding technique requires more care (Leduc and Bernhard 1967). GMA and HPMA are usually polymerized by irradiation with ultra-violet light (§5.2). A simple apparatus for this has been designed by Cole (1968). Methacrylates remain liquid at low temperatures and can therefore be used for dehydration and infiltration in the cold (§6.1).

5.6.2a *Glycol methacrylate*

Glycol methacrylate (GMA) is a clear, colourless liquid of low viscosity. It slowly polymerizes on storage and should be discarded if it is not readily miscible with water and a fresh lot obtained. It is not necessary to remove the inhibitor.

Blocks of pure GMA are very difficult to section and so *n*-butyl methacrylate (Leduc and Bernhard 1967) or styrene (Cope 1968) is added to the final embedding mixture to improve sectioning properties. Leduc and Bernhard (1967) recommended a mixture consisting of 7 parts of 97% GMA in water and 3 parts of *n*-butyl methacrylate (with inhibitor), containing 2%

Luperco (benzoyl peroxide paste) as catalyst, while Cope (1968) used GMA/
n-butyl methacrylate (7/3) or GMA/styrene (7/3) with 1.2% benzoyl peroxide
as catalyst.

Partially polymerized embedding medium is used for the final infiltration
and embedding and should have the viscosity of a thick syrup at 1–3 °C
(Leduc and Bernhard 1967).

Schedule for dehydration and embedding in GMA
(Cope and Williams 1968)

All procedures are carried out at 0–4 °C

80% GMA in water	15 min
100% GMA	4 changes, 15 min each
embedding medium + catalyst	1 hr
embedding medium + catalyst	1 hr
polymerization in partially	
polymerized embedding medium	48 hr under UV light at 20 °C

Gelatin and not polyethylene capsules should be used.

Glycol methacrylate can also be used for dehydration before embedding in
a polyester or epoxy resin (Kushida 1964b), and GMA can be copolymerized
with Epon or Vestopal (McGee-Russell and De Bruijn 1964) to yield blocks
with excellent sectioning properties.

5.6.2b   *Hydroxypropyl methacrylate*

Hydroxypropyl methacrylate (HPMA) is similar in appearance and properties to GMA; it should not be used unless it is completely miscible with
water in the ratio of about 4:1 at 20 °C (Leduc and Holt 1965). The miscibility
of water with HPMA *increases* as the temperature is lowered below 20 °C.
HPMA is polymerized by light, even in the absence of a catalyst, and should
be stored in the dark. The inhibitor need not be removed before use.

The embedding medium consists of 80% or 90% HPMA in water with
0.1% azonitrile as catalyst. Final infiltration and embedding is done in
partially polymerized methacrylate (§ 5.5.2).

Schedule for dehydration and embedding in HPMA
(Leduc and Holt 1965)

All procedures are carried out in the cold room using a mechanical shaker.

85% HPMA in H_2O	1 hr
85% HPMA in H_2O	1 hr

97% *or* 80% HPMA in H_2O 1 hr
97% *or* 80% HPMA in H_2O 1 hr
partially polymerized
embedding medium   1 hr
polymerization   12 hr-24 hr under UV at 10 °C
         or   2–3 days at 56 °C

Pease (1966a, b) used HPMA as an embedding medium following dehydration in ethylene glycol (§4.3.1), while Brinkley et al. (1967) used HPMA as an intermediate solvent for monolayers of cells grown in Falcon plastic flasks to be embedded in Epon 812. These flasks dissolve in propylene oxide (§5.8.2).

5.6.2c   *Embedding in the presence of water*

The first water-soluble methacrylate embedding medium proposed by Rosenberg et al. (1960) contained a small amount of water (about 3%), and Bernhard and his colleagues (e.g. Bartl and Bernhard 1966; Bernhard 1966) have shown that large proportions of water can be included in the final block. For example, Bartl and Bernhard (1966) embedded aldehyde-fixed tissue in a mixture of various water-soluble methacrylates (20%) with 0.28 M sucrose in water (80%). Such blocks are suitable for sectioning at low temperatures.

5.7   Other embedding media

The majority of the embedding media already described in this chapter are suitable for general use. A considerable number of other embedding media have been developed for electron microscopy, but none of them is yet sufficiently well established to warrant description in detail. The more promising of these are listed below. Most of them are water-soluble and can be polymerized in the presence of water to yield hydrophilic blocks suitable for cytochemical and immunochemical studies (see Lewis et al. 1974).

5.7.1   GELATIN

Gelatin has been tested as a water-soluble embedding medium in a number of laboratories (see Moretz et al. 1969b for references). A bacteriological grade of gelatin is used and the tissue is dehydrated by passing through gradually increasing concentrations of gelatin from 10% to 40% or 70% in water, the final solution containing 2% glycerin. The tissue in 40% or 70% gelatin is allowed to dry partially at 37 °C for some hours, and then small pieces of the gel are cut out and dried in a vacuum desiccator. There is a very large shrinkage on hardening and many authors have found sectioning to be difficult. Bernhard and Nancy (1964) embedded specimens in gelatin for sectioning at low temperatures (§6.4).

5.7.2 UREA-ALDEHYDE

Mixtures of urea and aldehydes can be polymerized in the presence of substantial quantities of water, providing an opportunity to preserve lipid structures in *situ* (Peterson and Pease 1970a). Casley-Smith (1967) embedded formaldehyde-fixed tissues in a commercial sample of urea-formaldehyde, with ammonium chloride as catalyst, while Peterson and Pease (1970b) used a mixture of 10 ml of 50% glutaraldehyde and 4 g of urea, with oxalic acid as catalyst. There are considerable difficulties in controlling the polymerization so as to obtain adequate infiltration of the embedding medium into the tissue, and the final blocks are hydrophilic and difficult to section.

5.7.3 PROTEIN-ALDEHYDE

Attempts have been made to cross-link other proteins, but there are considerable technical difficulties to overcome. Moretz et al. (1969b) tested agar-formaldehyde and found it to be less satisfactory than gelatin, while Robertson and Parsons (1970) reported that resorcinol-formaldehyde is not suitable for routine use.

Farrant and McLean (1969) used mixtures of serum and egg albumins with glutaraldehyde, but failed to obtain infiltration of the albumin into cells and organelles. The albumin was thus acting as an 'encapsulating' medium, (§3.6.4) rather than as an embedding medium. Nicholson (1971) embedded isolated chloroplasts in bovine serum albumin (BSA)-glutaraldehyde by a modification of the method of Farrant and McLean and obtained some promising results. He reported that there was some separation of the chloroplasts from the BSA matrix, suggesting incomplete infiltration of the BSA.

Kuhlmann and Viron (1972) describe a method of using albumin cross-linked with glutaraldehyde to 'embed' material which has not been dehydrated as a preliminary to the preparation of ultrathin frozen sections.

5.7.4 POLYAMPHOLYTES

McLean and Singer (1964) proposed the use of cross-linked polyampholytes as water-soluble embedding media for electron microscopy. Polyampholytes were prepared by the copolymerization of comparable molar quantities of two monomers of opposite sign to yield a block studded with more or less alternating negative and positive charges. The anionic monomer was methacrylic acid and the cationic monomer was dimethylaminoethyl methacrylate, used with a small amount of a cross-linking monomer tetramethylene dimethacrylate. Although this embedding medium appeared to be promising for immunochemical studies when it was first introduced, it has not been widely used and more attention is now being paid to the protein-aldehyde embedding media (§5.7.2 and §5.7.3).

5.8 Special embedding methods

5.8.1 FLAT EMBEDDING

In the standard embedding method (§5.2) gelatin or polyethylene capsules are used as embedding moulds. All ultramicrotomes (Reid 1974) are designed with chucks to hold the resultant cylindrical blocks. However, long specimens, such as fibres, flat specimens, such as monolayers of cells, and specimens which are to be sectioned in a particular orientation are more conveniently embedded in a rectangular block. Special 'vice-type' chucks are available on most ultramicrotomes to hold such blocks.

It is essential that the embedding medium can be easily separated from the mould and consequently moulds made out of polyethylene, aluminium foil or silicone rubber are used. Any shallow polyethylene containers are suitable, including mini-cube ice trays (Robertson et al. 1963), small disposable beakers cut off about 1 cm from the bottom (McCombs et al. 1968), small flat bottomed vials (EFFA mini-vials, see Appendix), and weighing trays (Fig. 5.6). McKinney and Walz (1969) recommend commercial polyethylene

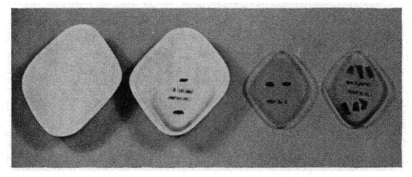

Fig. 5.6. Polyethylene weighing trays are suitable as moulds for flat embedding. The embedding medium easily separates from the tray after polymerization. A label is incorporated in the block near each specimen.

containers supplied by Nalge (see Appendix). These are 7/8 in in diameter and 1½-2 in deep, and have friction-fit covers, making them suitable for volatile embedding media such as the methacrylates. The height of the container is reduced to about 3/8 in by cutting with a razor blade. The hardened block is extruded from the mould by pushing with the thumbs, and the container is then reusable. If there is any difficulty in removing the block, the container is cut open.

When the exact orientation of the specimen in the block is important, rectangular moulds of the required dimensions can be constructed from polyethylene strips by the method described by Cheney and Ashhurst (1966).

Alternatively, moulds for flat embedding are made out of aluminium foil. A flat bottom to the mould is obtained by wrapping the foil around a metal block of the required dimensions and then pressing the block onto a piece of plate glass (Stetler 1972). Kushida (1969b) describes a technique in which a polyethylene container is lined with foil and covered with a sheet of cellophane for flat embedding with polyester resins. The foil protects the polyethylene which dissolves in the monomer resin.

Commercially available silicone rubber moulds and vinyl cups (Anderson and Doane 1967) are also suitable for flat embedding.

A number of specimens can be embedded in a single mould. A label (typed or indian ink) is incorporated in the block near each specimen (Fig. 5.6).

5.8.2   MONOLAYERS OF CELLS

Monolayers of cells are embedded in one of three ways:

(i)   The layer of cells is removed from the substrate before embedding and the cells are embedded in capsules by the standard method (§5.2).

(ii)   The layer of cells is removed from the substrate after embedding by separating the hardened block from the substrate.

(iii)   The layer of cells is grown on a substrate which can be sectioned and the cells are not separated from the substrate.

The third of these methods is preferable since the cells are easily damaged during the separation procedure. This damage is reduced if the cells are grown on glass coated with a special layer, or on various synthetic substrates, such as Falcon plastic Petri dishes (Table 5.14). Details of the preparation of these substrates can be found in the original references. The methods of cleaning glass before the deposition of a special layer, and the method of sterilizing the substrate before the cells are grown, are both important to ensure easy separation of the cells from the substrate at a later stage.

TABLE 5.14

Substrates to enable the separation of cell monolayers

glass–carbon	Robbins and Gonatas (1964), and others
glass–collagen	Heyner (1963)
glass–collodion	Meiselman et al. (1967); Flaxman et al. (1968)
glass–collodion–collagen	Descarries (1967)
glass–formvar	Storb et al. (1966)
glass–formvar–carbon	Breton-Gorius (1968)
glass–gelatin	Speirs and Turner (1966)
glass–silica	Kushida and Suzuki (1968)
glass–silicone	Gay (1955); Rosen (1962)
glass–Teflon	Chang (1971)
glass–tungsten oxide	Kushida and Suzuki (1970)
mica	Persijn and Scherft (1965); Yardley and Brown (1965); Kumagewa et al. (1968)
Nunclon petri dish	Biberfield (1968)
polystyrene (Falcon Petri dish)	Goldberg et al. (1963); Brinkley et al. (1967); Zagury et al. (1968); Ross (1972)
polypropylene	Zagury et al. (1968)
polyester (Melinex)	Firket (1966)
polyester–carbon	Clegg (1964)

5.8.2a   *Removal of cells from the substrate before embedding*
When the orientation of the cell layer is not important, it can be removed from the substrate by scraping with a rubber 'policeman' or a razor blade before embedding. The under surfaces of the cells are inevitably damaged during removal, but this damage can be reduced if the cells are fixed *in situ* before removal and if they are grown on a coverslip coated with a layer of collodion or carbon (e.g. Daniel et al. 1966). The break then tends to occur in this layer. Some cell damage always occurs during mechanical removal and it is preferable to remove the cells more gently. When a collodion layer is present, the collodion and cells can be floated off onto buffer after fixation, (Meiselman et al. 1967). Alternatively the monolayer is dehydrated in the usual way and the collodion detaches from the glass in absolute ethanol (Descarries 1967) or propylene oxide (Flaxman et al. 1968), releasing the cell layer. Cells grown on plastic Petri dishes are also released in propylene oxide with the help of slight agitation (Biberfield 1968), since these dishes dissolve in propylene oxide.

5.8.2b   *Separation of hardened blocks from the substrate; 'inverted capsule' technique*
Before the development of chucks on ultramicrotomes for holding flat blocks

it was customary to use the 'inverted capsule' technique to separate the hardened block containing the cells from the substrate. The monolayer of cells on the substrate is dehydrated and infiltrated with embedding medium in the same container in which it was fixed (§3.5). Propylene oxide is not used when the cells are grown in plastic Petri dishes because the dishes dissolve. Hydroxypropyl methacrylate is a convenient alternative intermediate solvent if one is required (Brinkley et al. 1967). The monolayer is then placed face-upwards in a plastic Petri dish and capsules filled to the brim with embedding medium are placed over selected regions of the monolayer. Individual cells can be selected by viewing the monolayer before (Barnicot and Huxley 1965; Lavail 1968) or after (Robbins and Gonatas 1964) fixation with a light microscope. A circular mark is made on the back of the substrate with a diamond-tipped slide marker or with glass marking ink (of a type not removed in the subsequent dehydration).

Ordinary gelatin capsules can be used directly. BEEM polyethylene capsules are modified by removing the cap and cutting off the tip (Lavail 1968; Branson 1971). The resultant cylinder is glued into position before being filled with embedding medium.

The embedding medium is then polymerized by heat or ultra-violet (UV) irradiation in the usual way (§5.2). When sheets of polyester film (Melinex) are used as a substrate, the application of heat is avoided since it may cause crystallization of the Melinex (Clegg 1964).

Various techniques have been tested for the separation of the hardened blocks from the substrate. The capsules can be removed from some substrates with a sharp pull after excess embedding medium has been trimmed away from the base of the capsule with a razor blade. These substrates include glass–silicone (Rosen 1962), glass–collagen (Heyner 1963), glass–carbon (Robbins and Gonatas 1964; Barnicot and Huxley 1965; Lavail 1968), polyester (Melinex) (Firket 1966), polystyrene (Falcon Petri dishes) (Zagury et al. 1968) and polypropylene (Zagury et al. 1968).

It is easier to remove methacrylate blocks than blocks of epoxy resins, and blocks of epoxy resins are more readily removed if this is done before the embedding medium had hardened completely. For example, partially polymerized blocks of Epon (12–16 hr at 60–65 °C) are easily separated from Falcon Petri dishes (Zagury et al. 1968) or glass (Branson 1971). The capsules should be pulled off immediately after removal from the oven while they are still warm. They are then returned to the oven to complete polymerization.

If it proves difficult to remove the capsules they can be loosened by cooling the substrate by placing it on solid CO_2 (dry-ice) (Howatson and Almeida

1958). The capsules should immediately be placed in a desiccator to prevent water condensing on the surface and damaging the cells which are only covered with a very thin layer of embedding medium. Damage is often observed and cooling should only be used as a last resort.

When mica substrates are used the mica cleaves on shearing leaving a thin layer of mica attached to the end of the block (Persijn and Scherft 1965; Yardley and Brown 1965). This layer is removed during sectioning.

5.8.2c *Separation of hardened blocks from the substrate; flat embedding technique*

When a vice-type chuck is available on the ultramicrotome for sectioning flat blocks, the monolayer on its substrate can be embedded in a flat embedding mould (§5.8.1). A thin layer of embedding medium is placed in the bottom of the mould and then the monolayer of cells on its substrate is placed face-downwards and covered with another thin layer of embedding medium. When the cells are grown in Falcon plastic Petri dishes, the dish, can be used as the embedding mould. Before polymerization most of the embedding medium is drained off leaving a thin layer covering the cells.

The resultant thin sheet containing the cells is separated from the substrate using one of the techniques used for capsules (§5.8.2b). The substrates that have proved suitable for removing sheets of cells without cooling include Falcon polystyrene dishes (Goldberg et al. 1963; Brinkley et al. 1967), polyester (Melinex)–carbon (Clegg 1964), glass–Teflon (Chang 1971) and glass–carbon (Langenberg et al. 1972).

Individual cells in monolayers growing on Melinex can be marked by inscribing a circle with a diamond scribe before fixation (Clegg 1964). The scribe indents the Melinex and the circle is visible on the surface of the hardened embedding medium after separation from the Melinex.

5.8.2d *Cells not separated from the substrate*

When the substrate can be sectioned it does not have to be removed from the embedded layer of cells. Suitable substrates for this embedding method are listed in Table 5.15; they are ideal for studying the relationships between the cells and their substrate (Fig. 5.7). The substrate and cell layer are cut into strips and then embedded in capsules or they are embedded flat as described in §5.8.2c. Millipore membrane filters, silicone rubber membranes, Saran Wrap and Falcon Petri dishes soften or even dissolve in propylene oxide and another intermediate solvent is used. Millipore membrane filters are

TABLE 5.15

Substrates that can be sectioned

Millipore membrane filters	Dalen and Nevalainen (1968)
	McCombs et al. (1968)
Nucleopore filters	Cornell (1969)
Saran Wrap	Holmes and Choppin (1968)
Silicone rubber membranes	Sharhar et al. (1972)
Methacrylate	Kjéllen et al. (1965)
Epon	Egeberg (1965)
	Zagury et al. (1966)
Araldite	Smith et al. (1969)
	Pegrum (1972)
Falcon plastic Petri dishes	Lipton and Konigsberg (1972)
	Nelson and Flaxman (1972)

composed of mixed esters of cellulose and are available with pores of various dimensions (Friedman et al. 1968) (see Appendix). They can be used to fix and embed cells in suspension (McCombs et al. 1968) (§3.6.5), as well as monolayers. Friedman et al. (1968) and England (1969) studied the fine structures of Millipore filters. Nucleopore filters (Cornell 1969) have the advantage that they have flatter surfaces than Millipore filters with relatively few pores (Fig. 5.8).

Coverslips made out of embedding media (methacrylate, Epon or Araldite) make excellent substrates, but care has to be taken to remove all toxic factors

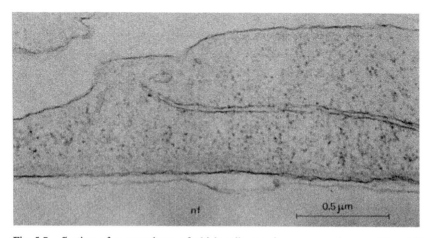

Fig. 5.7.   Section of a monolayer of chick cells growing on a nucleopore filter (nf). (Unpublished micrograph from a study by Audrey M. Glauert and Mary R. Daniel.)

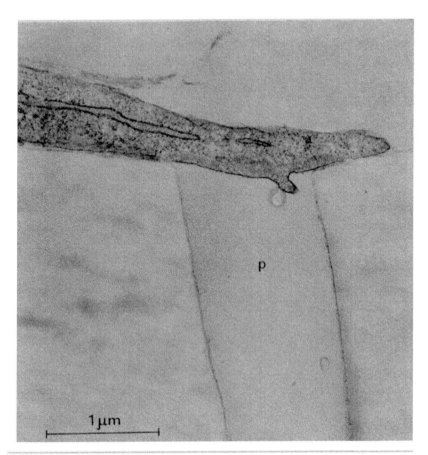

Fig. 5.8. A chick cell has grown over a pore (p) in a nucleopore filter. (Unpublished micrograph from a study by Audrey M. Glauert and Mary R. Daniel.)

before the cells are grown. The simplest way to make the coverslips is to polymerize a very thin layer of embedding medium in a container with a flat, level bottom, such as a bowl of flexible polyvinyl chloride (Egeberg 1965), or a polyethylene ice-cube container (Smith et al. 1969) or an aluminium foil cooking tray (Parker 1972). After hardening the lower surface of the embedding medium can be polished with a metal polish to provide a coverslip with good optical properties for light microscopy (Pegrum 1972). These coverslips are ideal for the selection of individual cells for sectioning since they are suitable for viewing at the highest magnification of the light microscope.

In two recent studies (Nelson and Flaxman 1972; Lipton and Konigsberg

1972) it was shown that it is possible to section Falcon plastic Petri dishes. A rectangular or wedge-shaped block containing a slice of the dish and the overlying cells embedded in epoxy resin is cut out and sectioned directly.

5.8.3 EMBEDDING METHODS FOR VERY SMALL QUANTITIES OF MATERIAL

The simplest method of embedding very small quantities of material is to carry out the whole preparative procedure in a capsule as described in (§3.8). The main problem with this method is to obtain adequate infiltration of the embedding medium into the specimen. An embedding medium of low viscosity should be chosen and the capsules warmed in an oven (50 °C–60 °C) before the final stage of centrifugation to reduce the viscosity of the embedding medium (Blackburn 1968).

The lid of a BEEM polyethylene capsule is a suitable mould for the flat embedding of single cells (Bondareff and Hydén 1969).

REFERENCES

Anderson, N. and F. W. Doane (1967), Epoxy embedding of thin-layer material in commercially available vinyl cups for light and electron microscopy, Stain technol. *42*, 169.
Anderson, W. A. and J. André (1968), The extraction of some cell components with pronase and pepsin from thin sections of tissue embedded in an Epon-Araldite mixture, J. Microscopie *7*, 343.
Argagnon, J. and L. Enjalbert (1964), Technique d'inclusion pour la microscopie électronique utilisant un polyester: le Rhodester 1108 CPSL, J. Microscopie *3*, 339.
Barnicot, N. A. and H. E. Huxley (1965), Electron microscope observations on mitotic chromosomes, Q. Jl Microsc. Sci. *106*, 197.
Bartl, P. and W. Bernhard (1966), Essais d'inclusion de tissus dans des gels plastiques fortement hydratés, J. Microscopie *5*, 51.
Bernhard, W. (1966), Progress and limitations of ultrastructural cytochemistry carried out on ultrathin sections, J. Histochem. Cytochem. *14*, 746.
Bernhard, W. and M.-T. Nancy (1964), Coupes à congélation ultrafines de tissu inclus dans la gélatine, J. Microscopie, *3*, 579.
Biberfield, P. (1968), A method for the study of monolayer cultures with preserved orientation and interrelationship, J. Ultrastruct. Res. *25*, 158.
Bird, A. F. (1964), Embedding and staining small nematodes for electron microscopy, Nature, Lond. *203*, 1300.
Blackburn, W. R. (1968), Carrier centrifuge tube for gelatin and polyethylene embedding capsules, J. Ultrastruct. Res. *23*, 362.
Bluemink, J. G. (1970), The first cleavage of the amphibian egg, J. Ultrastruct. Res. *32*, 142.
Bondareff, W. and H. Hydén (1969), Isolation and embedding of single nerve cells for electron microscopy, Stain technol. *44*, 87.
Branson, S. H. (1971), Epon-embedded cell monolayers, Expl. Cell Res. *65*, 253.
Breton-Gorius, J. (1968), Étude des leucocytes étalés par coupes sériées parallèles et perpendiculaires au plan d'étalement, J. Microscopie *7*, 95.

Brinkley, B. R., P. Murphy and L. C. Richardson (1967), Procedure for embedding *in situ* selected cells cultured *in vitro*, J. Cell Biol. *35*, 279.

Burke, C. N. and C. W. Geiselman (1971), Exact anhydride epoxy percentages for electron microscopy embedding (Epon), J. Ultrastruct. Res. *36*, 119.

Casley-Smith, J. R. (1967), Some observations on the electron microscopy of lipids, Jl R. microsc. Soc. *87*, 463.

Chang, J. P. (1971), A new technique for separation of coverglass substrate from epoxy-embedded specimens for electron microscopy, J. Ultrastruct. Res. *37*, 370.

Cheney, R. A. and D. E. Ashhurst (1966), A method for the orientation of tissues in epoxy-resin blocks, and a design for a rotating stage for holding blocks during trimming, Jl R. microsc. Soc. *86*, 441.

Clegg, M. D. (1964), A technique for the microbeam irradiation of single cells in tissue culture, Jl R. microsc. Soc. *83*, 433.

Cole, M. B. (1968), A simple apparatus for ultraviolet polymerization of water-soluble embedding media employed in electron microscopy, J. Microscopie *7*, 441.

Cope, G. H. (1968), Low-temperature embedding in water-miscible methacrylates after treatment with antifreezes, Jl R. microsc. Soc. *88*, 235.

Cope, G. H. and M. A. Williams (1968), Quantitative studies on neutral lipid preservation in electron microscopy, Jl R. microsc. Soc. *88*, 259.

Cope, G. H. and M. A. Williams (1969a), Quantitative studies on the preservation of choline and ethanolamine phosphatides during tissue preparation for electron microscopy. I. Glutaraldehyde, osmium tetroxide, Araldite methods, J. Microscopy *90*, 31.

Cope, G. H. and M. A. Williams (1969b), Quantitative studies on the preservation of choline and ethanolamine phosphatides during tissue preparation for electron microscopy. II. Other preparative methods, J. Microscopy *90*, 47.

Cornell, R. (1969), The use of nucleopore filters in ultrastructural studies of cell cultures, Expl Cell Res. *56*, 156.

Coulter, H. D. (1967), Rapid and improved methods for embedding biological tissues in Epon 812 and Araldite 502, J. Ultrastruct. Res. *20*, 346.

Craig, E. L., W. J. Frajola and M. H. Greider (1962), An embedding technique for electron microscopy using Epon 812, J. Cell Biol. *12*, 190.

Dalen, H. and T. J. Nevalainen (1968), Direct epoxy embedding for vertical sectioning of cells grown as a monolayer on Millipore filters, Stain technol. *43*, 217.

Daniel, M. R., J. T. Dingle, A. M. Glauert and J. A. Lucy (1966), The action of excess of vitamin A alcohol on the fine structure of rat dermal fibroblasts, J. Cell Biol. *30*, 465.

De Haller, G., C. F. Ehret and R. Naef (1961), Technique d'inclusion et d'ultramicrotomie, destinée à l'étude du développement des organelles dans une cellule isolée, Experientia *17*, 524.

DeLamater, E. D., E. Johnson, T. Schoen and C. Whitaker (1971), The use of styrenes as embedding media for electron microscopy, Proc. 29th Ann. Conf. EMSA, p. 488.

Descarries, L. (1967), An improved method for preparing cultivated nervous tissue for electron microscopic study, J. Microscopie *6*, 313.

Egeberg, J. (1965), Sandwich embedding of monolayers of cells in epoxy resin for ultrathin sectioning, Stain technol. *40*, 343.

England, M. A. (1969), Millipore filters studied in isolation and *in vitro* by transmission electron microscopy and stereoscanning electron microscopy, Expl Cell Res. *54*, 222.

Erlandson, R. A. (1964), A new Maraglas, D.E.R. 732, embedment for electron microscopy, J. Cell Biol. *22*, 704.

Estes, L. W. and J. V. Apicella (1969), A rapid embedding technique for electron microscopy, Lab. Invest. *20*, 159.

Farrant, J. L. and J. D. McLean (1969), Albumins as embedding media for electron microscopy, Proc. 27th Ann. Conf. EMSA, p. 422.

Finck, H. (1960), Epoxy resins in electron microscopy, J. biophys. biochem. Cytol. *7*, 27.

Firket, H. (1966), Polyester sheeting (Melinex O), a tissue culture support easily separable from epoxy resins after flat-face embedding, Stain technol. *41*, 189.

Flaxman, B. A., M. A. Lutzner and E. J. Van Scott (1968), Ultrastructure of cell attachment to substratum *in vitro*, J. Cell Biol. *36*, 406.

Freeman, J. A. and B. O. Spurlock (1962), A new epoxy embedment for electron microscopy, J. Cell Biol. *13*, 437.

Friedman, B., P. Blais and P. Shaffer (1968), Fine structure of millipore filters, J. Cell Biol. *39*, 208.

Gay, H. (1955), Serial sections of smears for electron microscopy, Stain technol. *30*, 239.

Gibbons, I. R. (1959), An embedding resin miscible with water for electron microscopy, Nature, Lond. *184*, 375.

Glauert, A. M. (1965), The fixation and embedding of biological specimens, in: Techniques for electron microscopy, D. H. Kay, ed. (Blackwell Scientific Publications, Oxford), p. 166.

Glauert, A. M. and R. H. Glauert (1958), Araldite as an embedding medium for electron microscopy, J. biophys. biochem. Cytol. *4*, 191.

Glauert, A. M., G. E. Rogers and R. H. Glauert (1956), A new embedding medium for electron microscopy, Nature, Lond. *178*, 803.

Goldberg, B., H. Green and G. J. Todaro (1963), Collagen formation *in vitro* by established mammalian cell lines, Expl Cell Res. *31*, 444.

Hayat, M. A. and R. Giaquinta (1970), Rapid fixation and embedding for electron microscopy, Tissue and Cell *2*, 191.

Heyner, S. (1963), *In situ* embedding of cultured cells or tissue, grown on glass, in epoxy resins for electron microscopy, Stain technol. *38*, 335.

Holmes, K. V. and P. W. Choppin (1968), On the role of microtubules in movement and alignment of nuclei in virus-induced syncytia, J. Cell Biol. *39*, 526.

Howatson, A. F. and J. D. Almeida (1958), A method for the study of cultured cells by thin sectioning and electron microscopy, J. biophys. biochem. Cytol. *4*, 115.

Jurand, A. and M. J. Ireland (1965), A slow rotary shaker for embedding in viscous media, Stain technol. *40*, 233.

Kellenberger, E., W. Schwab and A. Ryter (1956), L'utilisation d'un copolymère des polyesters comme matériel d'inclusion en ultramicrotomie, Experientia *12*, 421.

Kjellén, L., G. Lagermalm, A. Svedmyr and K. G. Thorsson (1965), Crystalline-like patterns in nuclei of cells infected with an animal virus, Nature, Lond. *175*, 505.

Korn, E. D. and R. A. Weisman (1966), Loss of lipids during preparation of amoebae for electron microscopy, Biochim. biophys. Acta *116*, 309.

Kuhlmann, W. D. and A. Viron (1972), Cross-linked albumin as supporting matrix in ultrathin cryo microtomy, J. Ultrastruct. Res. *41*, 385.

Kumagewa, M., M. Cattoni and G. C. Rose (1968), Electron microscopy of oral cells *in vitro*. III. *In situ* embedding of cultures in chambers of the circumfusion system, Tex. Rep. Biol. Med. *26*, 205.

Kurtz, S. M. (1961), A new method for embedding tissues in Vestopal W, J. Ultrastruct. Res. *5*, 468.

Kushida, H. (1959), On an epoxy resin embedding method for ultrathin sectioning, J. Electron Microscopy *8*, 72.

Kushida, H. (1960a), A new polyester embedding method for ultrathin sectioning, J. Electron Microscopy *9*, 113.

Kushida, H. (1960b), On the handling of epoxy resins and polyester resins for electron microscopy, J. Electron Microscopy *9*, 157.

Kushida, H. (1961a), A styrene-methacrylate resin embedding method for ultrathin sectioning, J. Electron Microscopy *10*, 16.

Kushida, H. (1961b), A new embedding method for ultrathin sectioning using a methacrylate resin with three dimensional polymer structure, J. Electron Microscopy *10*, 194.

Kushida, H. (1961c), On ultraviolet polymerization of polyester resins in embedding for electron microscopy, J. Electron Microscopy *10*, 201.

Kushida, H. (1962a), On ultraviolet polymerization of styrene resins in embedding for electron microscopy, J. Electron Microscopy *11*, 128.

Kushida, H. (1962b), A study of cellular swelling and shrinkage during fixation, dehydration and embedding in various standard media, J. Electron Microscopy *11*, 135.

Kushida, H. (1962c), Uranyl nitrate as a catalyst for ultraviolet polymerization in embedding, J. Electron Microscopy *11*, 253.

Kushida, H. (1963a), A modification of the water-miscible epoxy resin 'Durcupan' embedding method for ultrathin sectioning, J. Electron Microscopy *12*, 71.

Kushida, H. (1963b), An improved epoxy resin 'Epok 533', and polyethylene glycol 200 as a dehydrating agent, J. Electron Microscopy *12*, 167.

Kushida, H. (1964a), Improved methods for embedding with Durcupan, J. Electron Microscopy *13*, 139.

Kushida, H. (1964b), Glycol methacrylate as a dehydrating agent for embedding with polyester and epoxy resins, J. Electron Microscopy *13*, 200.

Kushida, H. (1965a), Durcupan as a dehydrating agent for embedding with polyester, styrene and methacrylate resins, J. Electron Microscopy *14*, 52.

Kushida, H. (1965b), Dehydration and embedding for electron microscopy. 2. Embedding, J. Electron Microscopy *14*, 251.

Kushida, H. (1965c), A new method for embedding with epoxy resin at room temperature, J. Electron Microscopy *14*, 275.

Kushida, H. (1966a), Further improved method for embedding with Durcupan, J. Electron Microscopy *15*, 95.

Kushida, H. (1966b), New embedding with D.E.R. 732 and Epon 812, J. Electron Microscopy *15*, 96.

Kushida, H. (1967), A new embedding method employing D.E.R. 736 and Epon 812, J. Electron Microscopy *16*, 278.

Kushida, H. (1969a), A new rotary shaker for fixation, dehydration and embedding, J. Electron Microscopy *18*, 137.

Kushida, H. (1969b), Flat embedding with polyester resins, J. Electron Microscopy *18*, 197.

Kushida, H. (1971), A new method for embedding with Epon 812, J. Electron Microscopy *20*, 206.

Kushida, H. and K. Fujita (1968), Methyl methacrylate as an auxiliary to infiltration for embedding with Vestopal W, J. Electron Microscopy *17*, 349.

Kushida, H. and K. Fujita (1971), New device for penetration of embedding media into specimens, J. Electron Microscopy *20*, 208.

Kushida, H. and K. Suzuki (1968), Evaporated silica as a stripping agent for coverslip-cultured cells embedded in epoxy resins, J. Electron Microscopy *17*, 350.

Kushida, H. and K. Suzuki (1970), Evaporated tungsten oxide as a stripping agent for monolayer-cultured cells embedded in epoxy resins, J. Electron Microscopy *19*, 191.

Langenberg, W. G., H. F. Schroeder, A. B. Welch and G. E. Cook (1972), Epoxy film separable from glass surfaces for selective light and electron microscopy of tissue and *in situ* grown cells, Stain technol. *47*, 303.

Lavail, M. M. (1968), A method of embedding selected areas of tissue cultures for electron microscopy, Tex. Rep. Biol. Med. *26*, 215.

Leduc, E. H. and W. Bernhard (1961), Ultrastructural cytochemistry. Enzyme and acid hydrolysis of nucleic acids and proteins, J. biophys. biochem. Cytol. *10*, 437.

Leduc, E. H. and W. Bernhard (1967), Recent modifications of the glycol methacrylate embedding procedure, J. Ultrastruct. Res. *19*, 196.

Leduc, E. H. and S. J. Holt (1965), Hydroxypropyl methacrylate, a new-water-miscible embedding medium for electron microscopy, J. Cell Biol. *26*, 137.

Leduc, E. H., V. Marinozzi and W. Bernhard (1963), The use of water-soluble glycol methacrylate in ultrastructural cytochemistry, Jl R. microsc. Soc. *81*, 119.

Lewis, P. R., D. P. Knight and M. A. Williams (1974), Staining methods for thin sections, in: Practical methods in electron microscopy, A. M. Glauert, ed. (North-Holland, Amsterdam).

Lipton, B. H. and I. R. Konigsberg (1972), A fine-structural analysis of the fusion of myogenic cells, J. Cell Biol. *53*, 348.

Lockwood, W. R. (1964), A reliable and easily sectioned epoxy resin embedding medium, Anat. Rec. *150*, 129.

Low, F. N. and M. R. Clevenger (1962), Polyester-methacrylate embedments for electron microscopy, J. Cell Biol. *12*, 615.

Luft, J. H. (1961), Improvements in epoxy resin embedding methods, J. biophys. biochem. Cytol. *9*, 409.

Maaløe, O. and A. Birch-Andersen (1956), On the organization of the 'nuclear material' in *Salmonella typhimurium*, Symp. Soc. gen. Microbiol. *6*, 261.

McCombs, R. M., M. Benyesh-Melnick and J. P. Brunschwig (1968), The use of Millipore filters in ultrastructural studies of cell cultures and viruses, J. Cell Biol. *36*, 231.

McGee-Russell, S. M. and W. C. De Bruijn (1964), Experiments on embedding media for electron microscopy, Q. Jl Microsc. Sci. *105*, 231.

McKinney, R. V. and E. J. Walz (1969), Flat embedding of thin tissues in commercially available polyethylene containers, Stain technol. *44*, 251.

McLean, J. D. and S. J. Singer (1964), Cross-linked polyampholytes. New water-soluble embedding media for electron microscopy, J. Cell Biol. *20*, 518.

Meiselman, N., A. Kohn and D. Danon (1967), Electron microscopic study of penetration of Newcastle disease virus into cells leading to formation of polykaryocytes, J. Cell Sci. *2*, 71.

Mohr, W. P. and E. C. Cocking (1968), A method of preparing highly vacuolated, senescent, or damaged plant tissue for ultrastructural study, J. Ultrastruct. Res. *21*, 171.

Mollenhauer, H. H. (1959), Permanganate fixation of plant cells, J. biophys. biochem. *6*, 431.

Mollenhauer, H. H. (1964), Plastic embedding mixtures for use in electron microscopy, Stain technol. *39*, 111.

Moretz, R. C., C. K. Akers and D. F. Parsons (1969a), Use of small angle X-ray diffraction to investigate disordering of membranes during preparation for electron microscopy. I. Osmium tetroxide and potassium permanganate, Biochim. biophys. Acta *193*, 1.

Moretz, R. C., C. K. Akers and D. F. Parsons (1969b), Use of small angle X-ray diffraction to investigate disordering of membranes during preparation for electron microscopy. II. Aldehydes, Biochim. biophys. Acta *193*, 12.

Nelson, B. K. and B. A. Flaxman (1972), *In situ* embedding and vertical sectioning for electron microscopy of tissue cultures grown on plastic Petri dishes, Stain technol. *47*, 261.

Newman, S. B., E. Borysko and M. Swerdlow (1949), New sectioning techniques for light and electron microscopy, Science *110*, 66.

Nicolson, G. L. (1971), Structure of the photosynthetic apparatus in protein-embedded chloroplasts, J. Cell Biol. *50*, 258.

Parker, R. A. (1972), Personal communication, Strangeways Research Laboratory, Cambridge.

Parsons, D. F. and E. B. Darden (1961), Optimal conditions for methacrylate embedding of certain tissues and cells sensitive to polymerization damage, Expl Cell Res. *24*, 466.

Pease, D. C. (1964), Histological techniques for electron microscopy, 2nd edition (Academic Press, New York and London).

Pease, D. C. (1966a), The preservation of unfixed cytological detail by dehydration with 'inert' agents, J. Ultrastruct. Res. *14*, 356.

Pease, D. C. (1966b), Anhydrous ultrathin sectioning and staining for electron microscopy, J. Ultrastruct. Res. *14*, 379.

Pegrum, S. M. (1972), Personal communication, University College, London.

Persijn, J. P. and J. P. Scherft (1965), Sheet mica – a nonadherent carrier for surface culture of cells to be embedded in Epon, Stain technol. *40*, 89.

Peterson, R. G. and D. C. Pease (1970a), Polymerizable glutaraldehyde-urea mixtures as water-soluble embedding media, Proc. 28th Ann. Conf. EMSA. p. 334.

Peterson, R. G. and D. C. Pease (1970b), Features of the fine structure of myelin embedded in water-containing aldehyde resins, Proc. 7th Int. Congr. Electron Microscopy, Grenoble, *1*, 409.

Rampley, R. D. and A. Morris (1972), A rapid method for polyester embedding, Proc. 5th. Eur. Reg. Conf. Electron Microscopy, Manchester, p. 224.

Reedy, M. K. (1965), Section staining for electron microscopy. Incompatibility of methyl nadic anhydride with permanganates, J. Cell Biol. *26*, 309.

Reid, N. (1974), Ultramicrotomy, in: Practical methods in electron microscopy, A. M. Glauert, ed. (North-Holland, Amsterdam).

Richardson, K. C., L. Jarett and E. H. Finke (1960), Embedding in epoxy resins for ultrathin sectioning in electron microscopy, Stain technol. *35*, 313.

Robbins, E. and N. K. Gonatas (1964), *In vitro* selection of the mitotic cell for subsequent electron microscopy, J. Cell Biol. *20*, 356.

Robbins, E. and G. Jentzsch (1967), Rapid embedding of cell culture monolayers and suspensions for electron microscopy, J. Histochem. Cytochem. *15*, 181.

Robertson, J. D., T. S. Bodenheimer and D. E. Stage (1963), The ultrastructure of Maunther cell synapses and nodes in goldfish brain, J. Cell Biol. *19*, 159.

Robertson, J. G. and D. F. Parsons, (1970), Myelin structure and retention of cholesterol in frog sciatic nerve embedded in a resorcinol-formaldehyde resin, Biochim. biophys. Acta *219*, 379.

Robson, E. A. (1964), The cuticle of *Peripatopsis moseleyi*, Q. Jl Microsc. Sci. *105*, 281.

Rosen, S. I. (1962), Cover-glass embedding in open-end capsules for electron microscopy, Stain technol. *37*, 195.

Rosenberg, M., P. Bartl and J. Leško (1960), Water-soluble methacrylate as an embedding medium for the preparation of ultrathin sections, J. Ultrastruct. Res. *4*, 298.

Ross, R. (1972), Personal communication, University of Washington, Seattle.

Ryter, A. and E. Kellenberger (1958a), Etude au microscope électronique de plasmas contenant de l'acide désoxyribonucléique, Z. Naturf. *13*, 597.

Ryter, A. and E. Kellenberger (1958b), L'inclusion au polyester pour l'ultramicrotomie, J. Ultrastruct. Res. *2*, 200.

Sharhar, A., O. Novick and Y. Straussman (1972), Utilization of silicon rubber membranes as supporting medium for cells for electron microscopy, Proc. 5th Eur. Reg. Conf. Electron Microscopy, Manchester, p. 226.

Shinagawa, Y. and Y. Uchida (1961), On the specimen damage of spinal cord due to polymerization of embedding media, J. Electron Microscopy *10*, 86.

Shinagawa, Y., S. Yahara and Y. Uchida (1962), Polymerization of epoxy resin for electron microscopy in the cold, J. Electron Microscopy *11*, 133.

Smith, W., E. W. Gray and J. M. K. Mackay (1969), A sandwich-embedding technique for monolayers of cells cultured on Araldite, J. Microscopy *89*, 359.

Speirs, R. S. and M. X. Turner (1966), Serial sampling of cell cultures using gelatin coated slides in coplin jars, Expl Cell Res. *44*, 661.

Spurlock, B. O., V. C. Kattine and J. A. Freeman (1963), Technical modifications in Maraglas embedding, J. Cell Biol. *17*, 203.

Spurr, A. R. (1969), A low-viscosity epoxy resin embedding medium for electron microscopy, J. Ultrastruct. Res. *26*, 31.

Stäubli, W. (1960), Nouvelle matière d'inclusion hydrosoluble pour la cytologie électronique, C. r. Séanc. Soc. Biol. *250*, 1137.

Stäubli, W. (1963), A new embedding technique for electron microscopy, combining a water-soluble epoxy resin (Durcupan) with water-insoluble Araldite, J. Cell Biol. *16*, 197.

Stein, O. and Y. Stein (1971), Light and electron microscopic radioautography of lipids: techniques and biological applications, Adv. Lipid Res. *9*, 1.

Steinbrecht, R. A. and K. D. Ernst (1967), Continuous penetration of delicate tissue specimens with embedding resin, Science Tools *14*, 24.

Stetler, D. A. (1972), Thin-layer embedding of filamentous plant parts for light and electron microscopy, Stain technol. *47*, 270.

Storb, R., R. L. Amy, R. K. Wertz, B. Fauconnier and M. Bessis (1966), An electron microscope study of vitally stained single cells irradiated with a ruby laser microbeam, J. Cell Biol. *31*, 11.

Szubinska, B. (1971), 'New membrane' formation in *Amoeba proteus* upon injury of individual cells, J. Cell Biol. *49*, 747.

Voelz, H. and M. Dworkin (1962), Fine structure of *Myxococcus xanthus* during morphogenesis, J. Bact. *84*, 943.

Ward, R. T. (1958), Prevention of polymerization damage in methacrylate embedding, J. Histochem. Cytochem. *6*, 398.

Watson, M. L. and W. G. Aldridge (1961), Methods for the use of indium as an electron stain for nucleic acids, J. biophys. biochem. Cytol. *11*, 257.

Winborn, W. B. (1963), Light and electron microscopy of the islets of Langerhans of the Saimiri monkey pancreas, Anat. Rec. *147*, 65.

Winborn, W. B. (1965), Dow epoxy resin with triallyl cyanurate, and similarly modified Araldite and Maraglas mixtures, as embedding media for electron microscopy, Stain technol. *40*, 227.

Yardley, J. and G. Brown (1965), Fibroblasts in tissue culture. Use of colloidal iron for ultrastructure localization of acid mucopolysaccharides. Lab. Invest. *14*, 501.

Zacks, S. I. (1963), Mechanical stirrer for epoxy resin embedding media, Stain technol. *38*, 60.

Zagury, D., G. D. Pappas and P. I. Marcus (1968), Preparation of cell monolayers for combined light and electron microscopy: staining in blocks, J. Microscopie *7*, 287.

Zagury, D., P. Zeitoun and M. Vielte (1966), Cytologie: méthode d'étude de cultures des cellules par la microscopie électronique, C. r. Acad. Sci. *262*, 1458.

Chapter 6

Low-temperature methods

Specimens are prepared at low temperatures with the aim of reducing the extraction of labile components by the fixative, dehydrating agent and embedding medium. In the simplest method a standard schedule is used but all the solutions are kept cold in an ordinary deep-freeze cabinet. The other low-temperature methods, freeze-drying, freeze-substitution and cryo-ultramicrotomy, are not sufficiently widely used to warrant detailed description here. They all involve an initial rapid freezing of the specimen and this is often the only means of fixation. Useful descriptions of the mechanism of freezing in biological systems are given by Bullivant (1970) and Rebhun (1972).

6.1 Fixation, dehydration and embedding at low temperatures

Fernández-Morán (1961) was one of the first to propose a low-temperature schedule for electron microscopy. Pieces of tissue were fixed at 0 °C, dehydrated in a graded series of acetone or methanol with the temperature progressively decreasing from 0 °C to − 25 °C, and then infiltrated with methacrylate at − 25 °C over a period of 3 to 5 days. The methacrylate was polymerized by ultra-violet (UV) irradiation (§ 5.2) at − 25 °C or at − 55 °C, the specimen temperature being below − 10 °C or − 40 °C respectively.

More recently Cope (1968) developed a method of low-temperature dehydration and embedding in water-miscible methacrylates (§ 5.6.2). The tissues were treated with a cryo-protective agent (dimethyl sulphoxide or dimethyl sulphoxide mixed with glycerol or ethylene glycol) and cooled to − 20 °C before fixation.

The following schedule was recommended:

(1) Cut up tissue in 10% antifreeze solution at 0 °C for 5 min.

(2) Transfer tissue to 25% antifreeze solution at 0 °C for 5 min.

(3) Transfer tissue to 50% antifreeze solution at 0 °C for 5 min.

(4) Cool tissue in 50% antifreeze solution slowly to − 20 °C by placing in a deep-freeze cabinet for 40 min.

(5) Fix tissue in 5% glutaraldehyde in 50% antifreeze solution at − 20 °C for 6 to 8 hr.

(6) Dehydrate tissue in 4 to 6 changes of 100% glycol methacrylate at − 20 °C for 45 min each.

(7) Infiltrate tissue with 2 changes of embedding monomer at − 20 °C for 45 min each.

(8) Place tissue in capsules in partially polymerized embedding medium at − 20 °C.

(9) Polymerize embedding medium by UV irradiation at − 10 °C to − 12 °C for 48 hr.

The embedding medium consisted of a mixture of glycol methacrylate and *n*-butyl methacrylate, or glycol methacrylate and styrene, with benzoyl peroxide as catalyst.

It was hoped that this low-temperature procedure would reduce extraction during dehydration and embedding, but Cope and Williams (1968) found that the retention of neutral glycerides was no better than with conventional procedures. However, the loss of phosphatidyl choline was reduced (Cope and Williams 1969).

6.2   Freeze-drying

In the freeze-drying procedure the ice is removed from the frozen specimen under vacuum. Considerable damage is caused when the freeze-dried tissue is subsequently infiltrated with the liquid embedding medium, and consequently this method of preparation has generally been superceded by freeze-substitution or cryo-ultramicrotomy. Details of the freeze-drying method are given by Sjöstrand (1967) and Rebhun (1972).

6.3   Freeze-substitution

The damage caused during the infiltration of dry tissues with the embedding medium can be avoided by using the method of freeze-substitution. Rapid freezing of the tissue is followed by solution ('substitution') of the ice at temperatures below 0 °C (Feder and Sidman 1958). The frozen tissue is kept in the substituting fluid for a long period (Table 6.1) so that the ice dissolves slowly without distorting the structure of the tissue. After removal of the ice the specimen is infiltrated with an embedding medium, often at low temperature. The embedding medium is polymerized in the cold by UV irradiation or by heat in the usual way.

The freeze-substitution methods used in electron microscopy are listed in Table 6.1. Reviews of the development of the technique have been published recently by Bullivant (1970) and Pease (1973), while full practical details are given by Rebhun (1972).

The most important step in the procedure is the initial cryofixation. Ice crystals form during freezing and it is common practice to use a cryoprotective agent, such as glycerol or dimethyl sulphoxide. It must be remembered, however, that these agents may themselves have an effect on the structure of the specimen (Rebhun 1972). For example, dimethyl sulphoxide induces a large increase in volume of the mitotic apparatus in living marine eggs (Rebhun and Sawada 1969). Ice crystal formation can also be reduced by partial dehydration of the tissue before freezing by exposing it to hypertonic conditions (Rebhun and Sander 1971).

The specimen should be cooled as rapidly as possible to keep ice crystal formation to a minimum. Fernández-Morán (1960, 1961) originally proposed using liquid helium II at − 272 °C, but the transfer of heat in liquid helium is poor due to the presence of a film of helium gas which surrounds the specimen (Bullivant 1965). Propane, propylene and Genetron 23 are the best coolants (Rebhun 1972), but Freon 22 is preferred because it is safer to use (Bullivant 1970), and can be easily stored in a deep freeze. It can be re-used after filtration at low temperature through filter paper (Rebhun 1972).

The specimen should be immersed in the coolant as rapidly as possible to prevent dehydration during transfer. Rebhun (1972) describes a simple device for shooting the specimen into the quenching fluid and recommends coating the specimen with a formvar film to reduce evaporation. Monroe et al. (1968) developed an original technique in which small tubular stainless steel projectiles were accelerated to supersonic velocities (400 m/sec) by the explosion of gunpowder. These projectiles passed through the tissue (left ventricle of the heart) cutting out a cylindrical biopsy on the way, and were then collected in a chamber containing liquid propane at − 188 °C (Fig. 6.1).

Rapid freezing can also be obtained by bringing the specimen into contact with the polished surface of a metal block which has previously been cooled to about − 200 °C by immersion in liquid nitrogen (Van Harreveld and

TABLE 6.1

Freeze-substitution methods in electron microscopy

Pre-treatment of tissue	Coolant	Substitution			Embedding		Reference
		Fluid	Temp.	Time	Medium	Polymerization	
Glycerinated	Helium II	$-272°C$	Alcohol-acetone-ethyl chloride	$-130°C$ to $-80°C$		Methacrylate $-100°C$ to $-75°C$	UV $-80°C$ to $-20°C$	Fernández-Morán (1960, 1961)
None	Helium II	$-272°C$	Methanol	$-75°C$	2 weeks	Methacrylate	UV $-5°C$	Bullivant (1960)
Glycerinated	Propane*	$-175°C$	Methanol	$-75°C$	2 weeks	Methacrylate	UV $-5°C$	Bullivant (1962)
Glycerinated	Propane*	$-175°C$	Ethanol	$-75°C$	2 weeks	Methacrylate or Durcupan or Epon	UV $-25°C$ Heat	Bullivant (1965)
None	Freon 12*	$-150°C$	1% OsO_4 in acetone	$-80°C$	3 days to 2 weeks	Araldite 502	Heat	Rebhun (1961)
Glycerinated	Propane*	$-170°C$ to $-160°C$	acetone-ethanol (1:1)	$-108°C$	2 weeks	Araldite 502	Heat	Rebhun and Gagné (1962)
Hypertonic sea water	Freon 12* or Freon 22*	$-155°C$ to $-150°C$	1% OsO_4 in acetone-ethanol (1:1)	$-100°C$	2 weeks	Epon	Heat	Rebhun and Sander (1971)

None	Isopentane*	− 160°C	GMA	− 60°C to − 40°C	2 weeks	GMA	UV − 50° to − 40°C	Bartl (1962)
None	Silver*	− 207°C	2% OsO₄ in acetone	− 85°C	2 to 3 days	Maraglas	Heat	Van Harreveld and Crowell (1964)
None or glycerinated	70% ethylene glycol or Brass* or Freon 22*	− 77°C − 196°C − 160°C	70% ethylene glycol in salt solution or Propylene glycol	− 50°C − 40°C	Few hr Few hr	HPMA		Pease (1967a, b)
DMSO	20% DMSO	− 20°C	Ethanol-25% glutaraldehyde	− 20°C	2 days	GMA	UV − 12°C to − 10°C	Cope (1968)
None	Propane*	− 188°C	Ethanol	− 75°C	2 weeks	Epon	Heat	Monroe et al. (1968)

Key

*	Cooled with liquid nitrogen
UV	Ultra-violet irradiation
GMA	Glycol methacrylate
HPMA	Hydroxypropyl methacrylate
OsO₄	Osmium tetroxide
DMSO	Dimethyl sulphoxide

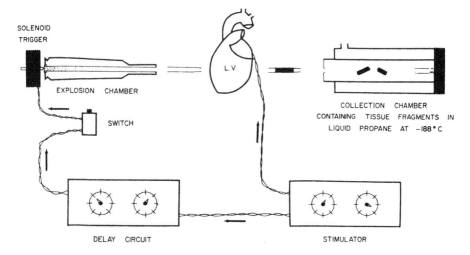

Fig. 6.1.   Schematic diagram of a ballistic technique for rapid transfer of a specimen into a coolant (from Monroe et al. 1968).

Crowell 1964; Pease 1967b). Even with the most rapid freezing methods only a narrow surface layer of a block of tissue is usually free of ice crystals (Van Harreveld and Crowell 1964).

After cryofixation the specimens are passed through the chosen substituting fluids, care being taken to ensure that the fluids are completely dry. Different fluids give very similar results (Rebhun and Sander 1971; Rebhun 1972). One of the main disadvantages of the freeze-substitution method is the long period required in the substituting fluid (Table 6.1). Periods of days or even weeks are required. Pease (1967a) used much shorter times (a few hours) for substitution with ethylene glycol, but it has been suggested by Bullivant (1970) that the tissue was allowed to warm up before substitution was complete. Pease (1973) has recently rejected this criticism.

The tissues may be stained by adding heavy metal salts to the substituting fluid (Fernández-Morán 1960, 1961; Rebhun 1961; Van Harreveld and Crowell 1964; Pease 1967a; Rebhun and Sander 1971), although no staining is observed with osmium tetroxide unless the fluid is allowed to warm up (Bullivant 1970). Alternatively, the sections are stained in the usual way.

The appearance of the tissue is very different from that observed after conventional fixation and embedding. In particular, membranes appear in negative contrast (Fig. 6.2). Even with the 'best' specimens only a small proportion of the cells is free of damage due to ice crystals.

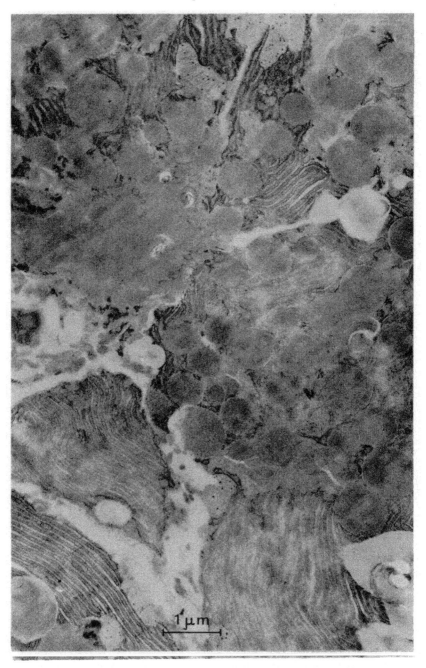

Fig. 6.2. Thin section of mouse pancreas prepared by freeze-substitution without chemical fixation. Embedded in Epon; section stained with lead hydroxide. (Unpublished micrograph from a study by Stanley Bullivant.)

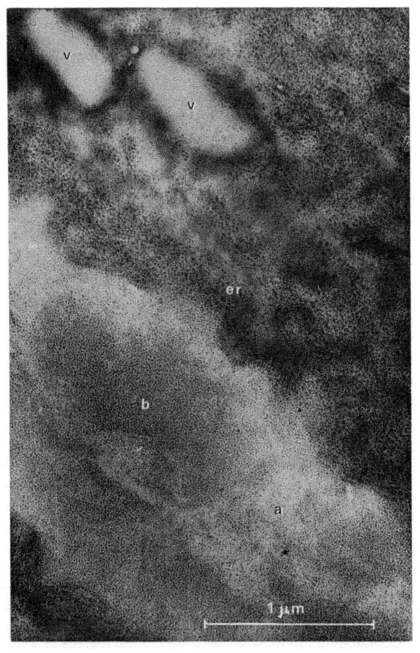

Fig. 6.3.   Ultra-thin dry frozen section of unfixed, unembedded mouse pancreas. The nucleus has light (a) and dense (b) areas. The ribosomes are visible on the rough endo-plasmic reticulum (er). The light areas (v) are secretory granules or vesicles.
(From Appleton 1972).

6.4 Cryo-ultramicrotomy

In recent years considerable advances have been made in the development of methods for cutting ultrathin sections of frozen specimens (Reid 1974) and special attachments for ultramicrotomes are commercially available. In some methods the tissue is fixed and then embedded in a water-soluble medium, such as gelatin, before freezing and sectioning (e.g. Bernhard and Leduc 1967), while in others the specimens are unfixed and unembedded (e.g. Appleton 1972; Fig. 6.3). This second, technically much more difficult method is necessary in studies on diffusible substances and promises to be of great value in cytochemistry.

The technique is being actively developed in a number of laboratories and it seems likely that before long it will become one of the established techniques in electron microscopy and warrant fuller description in later editions of this book.

REFERENCES

Appleton, T. C. (1972), 'Dry' ultra-thin frozen sections for electron microscopy and X-ray microanalysis: the cryostat approach, Micron *3*, 101.

Bartl, P. (1962), Freeze-substitution method using a water-miscible embedding medium, Proc. 5th Int. Congr. Electron Microscopy, Philadelphia *2*, P-4.

Bernhard, W. and E. H. Leduc (1967), Ultrathin frozen section. I. Methods and ultrastructural preservation, J. Cell Biol. *34*, 757.

Bullivant, S. (1960), The staining of thin sections of mouse pancreas prepared by the Fernández-Morán helium II freeze-substitution method, J. biophys. biochem. Cytol. *8*, 639.

Bullivant, S. (1962), Consideration of membranes and associated structures after cryofixation, Proc. 5th Int. Congr. Electron Microscopy, Philadelphia *2*, R-2.

Bullivant, S. (1965), Freeze substitution and supporting techniques, Lab. Invest. *14*, 1178.

Bullivant, S. (1970), Present status of freezing techniques, in: Some biological techniques in electron microscopy, D. F. Parsons, ed. (Academic Press, New York and London), p.101.

Cope, G. H. (1968), Low-temperature embedding in water-miscible methacrylates after treatment with antifreezes, Jl R. microsc. Soc. *88*, 235.

Cope, G. H. and M. A. Williams (1968), Quantitative studies on neutral lipid preservation in electron microscopy, Jl R. microsc. Soc. *88*, 259.

Cope, G. H. and M. A. Williams (1969), Quantitative studies on the preservation of choline and ethanolamine phosphatides during tissue preparation for electron microscopy. II. Other preparative methods, J. Microscopy *90*, 47.

Feder, N. and R. L. Sidman (1958), Methods and principles of fixation by freeze-substitution, J. biophys. biochem. Cytol. *4*, 593.

Fernández-Morán, H. (1960), Low-temperature preparation techniques for electron microscopy of biological specimens based on rapid freezing with liquid helium II, Ann. N.Y. Acad. Sci. *85*, 689.

Fernández-Morán, H. (1961), The fine structure of vertebrate and invertebrate photo-receptors as revealed by low-temperature electron microscopy, in: The structure of the eye, G. K. Smelser, ed. (Academic Press, New York), p. 521.

Monroe, R. G., W. J. Gamble, C. G. La Farge, R. Gamboa, C. L. Morgan, A. Rosenthal and S. Bullivant (1968), Myocardial ultrastructure in systole and diastole using ballistic cryofixation, J. Ultrastruct. Res. *22*, 22.

Pease, D. C. (1967a), Eutectic ethylene glycol and pure propylene glycol as substituting media for the dehydration of frozen tissue, J. Ultrastruct. Res. *21*, 75.

Pease, D. C. (1967b), The preservation of tissue fine structure during rapid freezing, J. Ultrastruct. Res. *21*, 98.

Pease, D. C. (1973), Substitution techniques, in: Advanced techniques in biological electron microscopy, J. K. Koehler, ed. (Springer-Verlag, Berlin), p. 35.

Rebhun, L. I. (1961), Applications of freeze-substitution to electron microscope studies of invertebrate oocytes, J. biophys. biochem. Cytol. *9*, 785.

Rebhun, L. I. (1972), Freeze-substitution and freeze-drying, in: Principles and techniques of electron microscopy. Biological applications, Vol. 2, M. A. Hayat, ed. (Van Nostrand Reinhold, New York), p. 1.

Rebhun, L. I. and H. T. Gagné (1962), Some aspects of freeze-substitution in electron microscopy, Proc. 5th Int. Congr. Electron Microscopy, Philadelphia *2*, L-2.

Rebhun, L. I. and G. Sander (1971), Electron microscope studies of frozen-substituted marine eggs. 1. Conditions for avoidance of intracellular ice crystallization, Am. J. Anat. *130*, 1.

Rebhun, L. I. and N. Sawada (1969), Augmentation and dispersion of the *in vivo* mitotic apparatus of living marine eggs, Protoplasma *68*, 1.

Reid, N. (1974), Ultramicrotomy, in: Practical methods in electron microscopy, A. M. Glauert, ed. (North-Holland, Amsterdam).

Sjöstrand, F. S. (1967), Electron microscopy of cells and tissues. Vol. 1. Instrumentation and techniques, (Academic Press, New York and London).

Van Harreveld, A. and J. Crowell (1964), Electron microscopy after rapid freezing on a metal surface and substitution fixation, Anat. Rec. *149*, 381.

Appendix

List of suppliers

Commercial suppliers of equipment
and materials for fixation, dehydration and embedding.

N.B.   The following list includes the names of suppliers known to the author at the present time; she will be interested to hear of names and addresses of other suppliers for inclusion in later editions.

General suppliers are listed first from whom the majority of the smaller items may be obtained. There follows a more comprehensive list with the items arranged in the same order as they are mentioned in the text.

★ Denotes a manufacturer

1. General suppliers of materials and equipment for electron microscopy

(a) **Agar Aids for Electron Microscopy**
(Agar Aids)
(U.K. agents for Ladd)
127a Rye Street
Bishop's Stortford
Hertfordshire
England

(b) **Ernest F. Fullam, Inc.** (EFFA)
P.O. Box 444
Schenectady
New York 12301
U.S.A.

Graticules Ltd.
(U.K. agents for EFFA)
Sovereign Way
Tonbridge
Kent
England

Touzart & Matignon
(French agents for EFFA)
3, Rue Amyot
75 Paris 5
France

(c) **EMscope Laboratories** (EMscope)
99, North Street
London SW4 OHQ
England

(d) **Ladd Research Industries Inc.** (Ladd)
(U.S. agents for Agar Aids)
P.O. Box 901
Burlington
Vermont 05401
U.S.A.

(e) **Polaron Equipment Ltd.** (Polaron)
(U.K. agents for Polysciences)
60/62, Greenhill Crescent
Holywell Estate
Watford
Hertfordshire
England

Ted Pella Co. (Pelco)
(U.S. Agents for Polaron)
P.O. Box 510
Tustin
California 92680
U.S.A.

(f) **Polysciences Inc.** (Polysciences)
(U.S. agents for Polaron)
Paul Valley Industrial Park
Warrington
Pennsylvania 18976
U.S.A.

(g) **Taab Laboratories** (Taab)
52, Kidmore End Road
Emmer Green
Reading
Berkshire
England

Extech International Corp. (Extech)
(U.S. agents for Taab)
177, State Street
Boston
Massachusetts 02109
U.S.A.

(h) **E. M. Zairyo-Shya**
308, Yotsuya Sango Building
2, Saemoncho Shinijuku-ku,
Tokyo
Japan

2. Buffer solutions (§2.2)

(a) *Phospate buffer*

Fisons Scientific Apparatus (Fisons)
Bishops Meadow Road
Loughborough
Leicestershire LE11 ORG
England

Polysciences   (*see* 1f)

(b) *Cacodylate buffer*

Polaron   (*see* 1e)
Taab   (*see* 1g)

(c) *Veronal-acetate buffer*

Ladd   (*see* 1d)
Polaron   (*see* 1e)
Polysciences   (*see* 1f)

(d) *Collidine buffer*

Ladd   (*see* 1d)
Polaron   (*see* 1e)
Polysciences   (*see* 1f)
Taab   (*see* 1g)

(e) *Buffer chemicals*

General suppliers   (*see* 1)
Fisons   (*see* 2a)

G. T. Gurr (Gurr)
Searle Scientific Services
Coronation Road
Cressex Industrial Estate
High Wycombe
Buckinghamshire
England

(f) *Purified collidine (distilled)*

Eastman Kodak
Distillation Products
New York, N.Y.
U.S.A.

3. Fixatives (Chapter 2)

(a) *Osmium tetroxide* (§ 2.3)

General Suppliers	(*see* 1)
Fisons	(*see* 2a)
Gurr	(*see* 2e)

Johnson, Matthey Chemicals Ltd.
73–83, Hatton Garden
London EC1
England

Note: Osmium tetroxide may be returned to Johnson, Matthey Chemicals for regeneration and re-use.

(b) *Glutaraldehyde* (§ 2.4)

General suppliers	(*see* 1)

Fisher Scientific (Fisher)
633, Greenwich Street
New York, N.Y. 10014
U.S.A.

Gurr	(*see* 2e)

Kodak Ltd.
(U.K. agents for Fisher)
Chemical Division
Kirkby
Liverpool
England

Union Carbide Corp.
270, Park Avenue
New York, N.Y. 10017
U.S.A.

(c) *Formaldehyde* (§ 2.5)

EFFA	(*see* 1b)
EMscope	(*see* 1c)
Polaron	(*see* 1e)
Polysciences	(*see* 1f)
Taab	(*see* 1g)
Fisons	(*see* 2a)
Gurr	(*see* 2e)

(d) *Paraformaldehyde* (§ 2.5)

EMscope	(*see* 1c)
Ladd	(*see* 1d)
Polaron	(*see* 1e)
Taab	(*see* 1g)
Fisons	(*see* 2a)
Gurr	(*see* 2e)

(e) *Acrolein* (acrylaldehyde) (§ 2.6)

General suppliers	(*see* 1)
Fisons	(*see* 2a)

Shell Chemicals Ltd.
Shell House
Downstream Buildings
London SE1
England

or:

Shell Chemical Corp.
Industrial Chemical Division
415 Madison Avenue
New York, N.Y.
U.S.A.

(f) *Other aldehydes* (§ 2.6)

EMscope	(*see* 1c)
Polaron	(*see* 1e)
Polysciences	(*see* 1f)
Taab	(*see* 1g)
Fisons	(*see* 2a)
Gurr	(*see* 2e)

(g) *Potassium permanganate* (§ 2.7)

General suppliers	(*see* 1)
Fisons	(*see* 2a)
Gurr	(*see* 2e)

(h) *Potassium ferricyanide* (§ 2.8)

EMscope	(*see* 1c)
Fisons	(*see* 2a)
Taab	(*see* 1g)

(i) *Uranyl acetate* (§ 2.9)

General suppliers	(*see* 1)
Fisons	(*see* 2a)
Gurr	(*see* 2e)

4. Equipment for fixation (Chapter 3)

(a) *Glass vials*, with plastic caps (§ 3.1)

Agar Aids	(*see* 1a)
EFFA	(*see* 1b)
EMscope	(*see* 1c)
Polaron	(*see* 1e)

(b) *Tweezers* (§ 3.1)

General suppliers	(*see* 1)

Mason & Morton Ltd.
Fir Tree House
Headstone Drive
Wealdstone, Harrow
Middlesex HA3 5QS
England

(c) *Dental wax* (§ 3.3)

EFFA	(*see* 1b)
EMscope	(*see* 1c)
Ladd	(*see* 1d)
Polaron	(*see* 1e)
Taab	(*see* 1g)

(d) *Razor blades*, single-edged (§ 3.3)

EFFA	(*see* 1b)
EMscope	(*see* 1c)
Polaron	(*see* 1e)
Taab	(*see* 1g)

(e) *Millipore membrane filters* (§ 3.5)

Millipore Filter Corp.
Millipore House
Abbey Road
London NW10 7SP
England

(f) *Microfuge* (§ 3.8)

Beckman-RIIC Ltd.
Eastfield Industrial Estate
Glenrothes
Fife
Scotland

(g) *Centrifuge tubes modified to hold BEEM capsules* (§ 3.8)

EFFA	(*see* 1b)

5. Additives for fixation by perfusion (§ 3.3)

(a) *Dextran*

Fisons	(*see* 2a)
Gurr	(*see* 2e)
Polysciences	(*see* 1f)
Taab	(*see* 1g)

(b) *Gum acacia*

Fisons	(*see* 2a)

(c) *Polyvinylpyrrolidone* (*PVP*)

Fisons	(*see* 2a)
Gurr	(*see* 2e)
Polysciences	(*see* 1f)
Taab	(*see* 1g)

6. Encapsulating media (§ 3.6)

(a) *Agar*

Fisons	(*see* 2a)
Gurr	(*see* 2e)
Polysciences	(*see* 1f)
Taab	(*see* 1g)

(c) *Bovine serum albumin* (BSA)

EFFA	(*see* 1b)
Gurr	(*see* 2e)
Polysciences	(*see* 1f)
Taab	(*see* 1g)

(b) *Fibrinogen and thrombin*

Parke Davis and Co. Ltd.
Hounslow
Middlesex
England

(d) *Egg albumen*

Fisons	(*see* 2a)

7. Dehydrating agents and intermediate solvents (Chapter 4)

(a) *Ethylene glycol (ethanediol)*

Fisons	(*see* 2a)

(b) *Polyethylene glycol 200 (Carbowax 200)*

Fisons	(*see* 2a)
Gurr	(*see* 2e)
Polysciences	(*see* 1f)
Taab	(*see* 1g)
Union Carbide	(*see* 3b)

(c) *Propylene oxide (1,2-epoxy propane)*

General suppliers	(*see* 1)
Fisons	(*see* 2a)
Gurr	(*see* 2e)

(d) *Styrene*

Polaron	(*see* 1e)
Polysciences	(*see* 1f)
Taab	(*see* 1g)

8. Rotary shakers and mixers (§ 5.2)

(a) *Epoxy mixer*

EFFA	(*see* 1b)

(b) *Jurand and Ireland Rotary Shaker*

EFFA	(*see* 1b)

(c) *Penetron Rotary Shaker (Kushida design)*

Polysciences	(*see* 1f)

(d) *Rotamixers*

EMscope★	(*see* 1c)

(e) *Rotary shakers and mixers VEM-16*

Sakura Finetechnical Co Ltd.★
9 Honcho 3-chome, Nihombashi
Chuo-ku
Tokyo 103
Japan

(f) *Sunkay 'Swirling' Shaker*

Polaron	(*see* 1e)

(g) *Taab Rotator*

Taab	(*see* 1g)

(h) *Zacks Teflon Stirrer*

Polaron   (*see* 1e)
Polysciences   (*see* 1f)

9. Epoxy resin embedding components (§ 5.3)

(a) *Table A.1*

R. P. Cargille Laboratories Inc.
Cedar Grove
New Jersey 07009
U.S.A.

Ciba-Geigy (Ciba U.K.)
Duxford
Cambridge
England

Ciba Co., Inc. (Ciba U.S.)
Plastics Division
Kimberton
Pennsylvania
U.S.A.

and

Ciba Products Corp.
Fair Lawn
New Jersey
U.S.A.

Dow Chemical Co.
Midland
Michigan 48640
U.S.A.

and

Dow Chemical Europe, S.A.
Alfred Escher Strasse 39
8207 Zurich
Switzerland

Fluka AG
Chemische Fabrik
Buchs SG
Switzerland

(b) *Epok 533*

Oken Shoji Co.★
311, Kobikikan 7
6-chome
Ginza-higashi
Chuo-ku
Tokyo
Japan

(c) *Maraglas 655*

Marblette Corp★
Long Island City
New York
U.S.A.

(d) *DMP-30*

Rohm & Haas Co.
Washington Square
Philadelphia
Pennsylvania
U.S.A.

(e) *BDMA (Benzyl dimethyl amine)*

Maumee Chemical Co.
2, Oak Street
Toledo
Ohio
U.S.A.

(f) *DMAE (Dimethylaminoethanol) (S-1)*

Pennsalt Chemical Corp.
Three Penn. Center
Philadelphia
Pennsylvania 19102
U.S.A.

TABLE A.1

Suppliers of epoxy resin embedding components

Key: ○ Manufacturer
w-s water soluble

	Cargille	Ciba (U.K.)	Ciba (U.S.)	Dow	EFFA (see 1b)	EMscope (see 1c)	Fluka	Gurr (see 2e)	Ladd (see 1d)	Polaron (see 1e)	Polysciences (see 1f)	Shell (see 2a)	Taab (see 1g)	Union Carbide (see 3b)	Zairyo-Shya (see 1h)
Embedding kits															
Araldite CY 212 or	×	—	—	—	—	×	—	—	—	×	×	—	×	—	—
Durcupan ACM	—	—	—	—	—	×	×	×	—	×	—	—	×	—	—
Araldite 502	×	—	—	—	—	—	—	—	×	—	—	—	—	—	—
Araldite 506	—	—	—	×	—	—	—	—	—	—	—	—	—	—	—
Araldite 6005	—	—	—	—	×	—	—	—	×	—	—	—	—	—	—
DER 332	—	—	—	—	—	—	—	—	—	×	—	—	—	—	—
Epon 812	×	—	—	—	×	×	—	—	×	×	×	×	×	—	—
Maraglas 655	×	—	—	—	×	—	—	—	×	×	×	—	—	—	—
ERL 4206	—	—	—	—	×	×	—	—	×	×	×	—	×	—	×
Durcupan (w–s)	—	—	—	—	—	×	×	×	—	×	×	—	×	—	—
Epoxy resins															
Araldite CY 212 or	—	○	—	—	—	×	—	—	—	—	—	—	×	—	—
Durcupan ACM	—	—	—	—	—	×	×	×	—	×	—	—	×	—	—
Araldite 502	×	—	○	—	×	—	—	—	—	×	×	×	—	—	—
Araldite 506	×	—	○	—	×	—	—	—	—	×	×	—	—	—	—
Araldite 6005	×	—	○	—	×	—	—	—	—	×	×	—	—	—	—
DER 332	—	—	—	○	—	—	—	—	—	×	×	—	—	—	×
DER 334	—	—	—	○	—	—	—	—	—	—	—	—	—	—	—
ERL 4206	—	—	—	—	—	×	—	—	—	×	×	—	×	○	×
Epon 812	—	—	—	—	×	×	—	×	×	×	×	○	×	—	×
Epon 815	—	—	—	—	—	—	—	—	—	×	—	○	—	—	×
Maraglas 655	—	—	—	—	×	—	—	—	×	×	×	—	—	—	—
Aquon (w-s)	—	—	—	—	—	—	—	—	—	×	—	—	—	—	—
Durcupan (w-s)	—	—	○	—	—	×	×	×	—	×	—	—	×	—	×
Hardeners															
DDSA	—	×	—	—	×	×	×	×	×	×	×	—	×	—	×
HHPA	—	—	—	—	—	—	—	×	—	×	×	—	×	—	—
MNA or NMA	—	—	—	—	×	×	—	×	×	×	×	—	×	—	×
NSA	—	—	—	—	—	×	—	—	—	×	×	—	×	—	×
Accelerators															
BDMA	—	—	—	—	×	×	—	×	×	×	×	—	×	—	—
DMAE (S–1)	—	—	—	—	—	×	—	—	—	×	×	—	×	—	—
DMP-30	—	×	×	—	×	×	×	×	×	×	×	—	×	—	—
Plasticizers and flexibilizers															
Cardolite NC-513	—	—	—	—	×	—	—	×	×	×	×	—	—	—	—
DER 732	—	—	—	○	—	—	—	—	—	×	×	—	—	—	×
DER 736	—	—	—	○	—	×	—	—	×	×	×	—	×	×	×
Dibutyl phthalate	—	×	×	—	×	×	×	×	—	×	×	—	×	—	—
Triallyl cyanurate	—	—	—	—	—	—	—	—	—	—	×	—	—	×	—

Fabriek van Chemische Producten
(European agents for Pennsalt)
Vondelingenplaat N.V.
P.O. Box 7120
Rotterdam
The Netherlands

(g) *Thiokol LP-8*

Thiokol Chemical Corp.★
Trenton
New Jersey
U.S.A.

10. Polyester resin embedding kits (§ 5.4)

(a) *Selectron*

Pittsburgh Plate Glass Co.★
Plastic Sales
Paint and Brush Division
Pittsburgh
Pennsylvania
U.S.A.

Polysciences   (*see* 1f)

(b) *Vestopal W*

Polaron   (*see* 1e)
Polysciences   (*see* 1f)

11. Polyester resins (§ 5.4)

(a) *Beetle 4116 and 4134*

B.I.P. Chemicals Ltd.★
Oldbury
Birmingham
England

(b) *Rhodester 1108*

Société des Usines Chimiques★
Rhône-Poulenc
France

(c) *Rigolac 2004 and 70F*

Riken Goseijushi Co.★
3, 6-chome
Ginza 6-3
Chuo-ku
Tokyo
Japan

(d) *Selectron*

Pittsburgh Plate Glass Co.   (*see* 10a)
Polysciences   (*see* 1f)

(e) *Vestopal W*

Martin Jaeger★
Vésenaz
Geneva
Switzerland

Gurr   (*see* 2e)
Polaron   (*see* 1e)
Polysciences   (*see* 1f)
Taab   (*see* 1g)

12. Initiators and accelerators for polyester resins and methacrylates (§ 5.4 and § 5.5)

(a) *Azo-bis-iso-butyronitrile*

Eastman Organic Chemicals
Rochester 3
New York 14650
U.S.A.

Polysciences   (*see* 1f)
Taab   (*see* 1g)

(b) *Benzoin*

Fisons	(*see* 2a)
Polysciences	(*see* 1f)
Taab	(*see* 1g)

(c) *Benzoyl peroxide*

Fisons	(*see* 2a)
Gurr	(*see* 2e)
Ladd	(*see* 1d)
Polysciences	(*see* 1f)
Taab	(*see* 1g)
Polaron	(*see* 1e)

(d) *Benzoyl peroxide paste (Luperco)*

Gurr	(*see* 2e)
Polysciences	(*see* 1f)
Riken Goseijushi★	(*see* 11c)
Taab	(*see* 1g)

Wallace and Tiernan Inc.
Lucido Division
Buffalo
New York 14240
U.S.A.

(e) *Cobalt naphthenate*

Gurr	(*see* 2e)
Jaeger	(*see* 11e)
Polysciences	(*see* 1f)
Taab	(*see* 1g)

(f) *Divinyl benzene*

Dow	(*see* 9a)
Polysciences	(*see* 1f)
Taab	(*see* 1g)

(g) *t-butyl perbenzoate*

Gurr	(*see* 2e)
Jaeger	(*see* 11e)
Taab	(*see* 1g)
Wallace and Tiernan	(*see* 12d)

(h) *Uranyl nitrate*

Fisons	(*see* 2a)
Gurr	(*see* 2e)
EMscope	(*see* 1c)
Ladd	(see 1d)
Taab	(*see* 1g)
Polaron	(*see* 1e)

13. Methacrylate embedding kits (§5.5)

(a) *n-butyl/methyl methacrylate*

Ladd	(*see* 1d)
Polaron	(*see* 1e)

(b) *Glycol methacrylate (GMA) (Water-soluble)*

Polaron	(*see* 1e)
Polysciences	(*see* 1f)
Zairyo-Shya	(*see* 1h)

(c) *HPMA (hydroxypropyl methacrylate)*

Polaron	(*see* 1e)
Taab	(*see* 1g)
Zairyo-Shya	(*see* 1h)

14. Methacrylates (§5.5)

(a) *n-butyl and methyl methacrylates*

Gurr	(*see* 2e)
Ladd	(*see* 1d)
Polaron	(*see* 1e)
Polysciences	(*see* 1f)
Taab	(*see* 1g)

(b) *Glycol methacrylate (GMA) (2-hydroxy ethyl methacrylate)*

Eastman Kodak Co.★
343, State Street
Rochester
New York 14650
U.S.A.

(b) *Glycol methacrylate (continued)*

Kodak Ltd.	(*see* 3b)
Polaron	(*see* 1e)
Polysciences	(*see* 1f)
Rohm & Haas	(*see* 9d)
Taab	(*see* 1g)

(c) *HPMA (2-hydroxy propyl methacrylate)*

Polaron	(*see* 1e)
Polysciences	(*see* 1f)
Rohm & Haas	(*see* 9d)
Taab	(*see* 1g)

(d) *Other methacrylates*

Polysciences	(*see* 1f)
Taab	(*see* 1g)

15. Capsules (Chapter 5)

(a) *Gelatin capsules*

General Suppliers	(*see* 1)
Parke Davis	(*see* 6b)

(c) *Taab polyethylene and polypropylene capsules*

Taab	(*see* 1g)

(d) *Polaron 'peelcap' capsules*

Polaron ★	(*see* 1e)

(b) *BEEM polyethylene capsules*

**Better Equipment for Electron
Microscopy Inc. (BEEM)** ★
P.O. Box 132
Jerome Avenue Station
Bronx
New York 10468
U.S.A.

General suppliers	(*see* 1)

(e) *BEEM Bojax embedding blocks
(sets of plastic capsules)*

BEEM ★	(*see* 15b)
Ladd	(*see* 1d)
Pelco	(*see* 1e)

(f) *Micromoulds (sets of plastic capsules)*

EFFA	(*see* 1b)
Pelco	(*see* 1e)

16. BEEM capsule holders (§ 5.2)

BEEM ★	(*see* 15b)
EFFA	(*see* 1b)
EMscope	(*see* 1c)
Ladd	(*see* 1d)
Polaron	(*see* 1e)

17. Embedding moulds for flat embedding (§ 5.8)

(a) *EFFA mini-vials*

EFFA ★	(*see* 1b)

(b) *Polyethylene moulds*

Nalge Co. Inc. ★
Rochester
New York
U.S.A.

(c) *Silicone rubber moulds*

EFFA	(*see* 1b)
Ladd	(*see* 1d)
Polaron	(*see* 1e)

C. W. French Inc.
58, Bittersweet Lane
Weston
Massachusetts 02193
U.S.A.

(d) *Vinyl cups*

Fabri-Kal Co.★
Kalamazoo
Michigan
U.S.A.

18. Ultra-violet lamps (§5.2)

(a) *UV curing chamber*

| Ladd | (*see* 1d) |

(b) *UV lamps*

EFFA	(*see* 1b)
Polysciences	(*see* 1f)
Taab	(*see* 1g)
Polaron	(*see* 1e)

P. W. Allen and Co.
253, Liverpool Road
London W1
England

19. Substrates for cell monolayers (§3.5 and §5.8)

(a) *Falcon Petri dishes (polystyrene)*

Falcon Plastics Co.★
Culver City
California
U.S.A.

Bio-Cult Laboratories Ltd.
3, Washington Road
Abbotsinch Industrial Estate
Paisley
Scotland

(b) *Nunclon Petri dishes*

Nunc A/S★
Roskilde
Denmark

James A. Jobling and Co. Ltd.
(U.K. agents for Nunc A/S)
Stone
Staffordshire ST15 OBG
England

(c) *Melinex film (polyester)*

Boyden Data Papers, Ltd.
Trade Service Division
Parkhouse Street
Camberwell
London SE5
England

(d) *Mica*

Agar Aids	(*see* 1a)
EFFA	(*see* 1b)
EMscope	(*see* 1c)
Polaron	(*see* 1e)

(e) *Teflon-coated coverslips*

| Polysciences | (*see* 1f) |

(f) *Nucleopore filters*

General Electric Company
U.S.A.

20. Dimethyl sulphoxide (*cryo-protective agent*) (§6.1)

EMscope	(*see* 1c)
Fisons	(*see* 2a)
Polysciences	(*see* 1f)
Taab	(*see* 1g)

21. Coolants (§6.3)

(a) *Freon 12 (dichlorodifluoromethane) and Freon 22 (monochlorofluoromethane)*

Under trade name '*Can-o-gas*'

Balzers AG
FL 9496 Blazers
Liechtenstein

and

Balzers High Vacuum Ltd
Northbridge Road
Berkhamsted
Hertfordshire
England

and

Balzers High Vacuum Corporation
PO Box 10816
Santa Ana
California 92711
U.S.A.

Dean and Wood (London) Ltd
83/85 Mansell Street,
London, E1
England

Under trade name '*Halocarbon 12 and 22*'

Air Products
Speciality Gases Department
Stenor Street
Cobridge, Burslem
Stoke-on-Trent
England

Under trade name '*Ucon*' (*Freon 22*)

J. T. Baker Chemical Company
Phillipsburg
New Jersey
U.S.A

Under trade name '*Arcton*'

I.C.I. Ltd
Mond Division
Rocksavage Works
Runcorn
Cheshire
England

(b) *Isopentane*

BDH Chemicals Ltd
Poole,
Dorset, BH12 4NN
England

(c) *Propane*

BDH (see 21b)

Index for list of suppliers

Numbers referred to are the sections in the preceding list of addresses

Accelerators, for epoxy resins, 9, Table A.1
  for methacrylates, 12
  for polyester resins, 12
Acrolein, 3(e)
Agar, 6(a)
Aldehydes, 3(f)
Araldite CY 212, 502, 506 and 6005,
  Table A.1
Azo-bis-iso-butyronitrile, 12(a)

BEEM capsules, 15(b)
  centrifuge tubes for, 4(g)
  holders for, 16
BEEM embedding blocks (Bojax), 15(e)
Beetle 4116 and 4134, 11(a)
Benzoin, 12(b)
Benzoyl, peroxide, 12(c)
  paste, 12(d)
Benzyl dimethyl amine (BDMA), 9(e),
  Table A.1
Bovine serum albumin (BSA), 6(c)
Buffer, cacodylate, 2(b)
  chemicals, 2(e)
  collidine, 2(d)
  phosphate, 2(a)
  veronal-acetate, 2(c)

Cacodylate buffer, 2(b)
Capsules, 15
  BEEM, 15(b)
  gelatin, 15(a)
  Polaron, 15(d)
  Taab, 15(c)
Capsule holders, BEEM, 16
Carbowax 200, 7(b)

Cardolite NC-513, Table A.1
Centrifuge tubes for BEEM capsules, 4(g)
Cobalt naphthenate, 12(e)
Collidine buffer, 2(d)
Collidine, purified, 2(f)
Coolants, 21
Coverslips, Teflon-coated, 19(e)

Dehydrating agents, 7
Dental wax, 4(c)
DER 332 and 334, Table A.1
DER 732 and 736, Table A.1
Dextran, 5(a)
Dibutyl phthalate, Table A.1
Dimethyl aminoethanol (DMAE),
  9(f), Table A.1
Dimethyl sulphoxide, (20)
Divinyl benzene, 12(f)
DMP-30, 9(d), Table A.1
Durcupan, water-soluble, Table A.1
Durcupan ACM, Table A.1

EFFA mini-vials, 17(a)
Egg albumen, 6(d)
Embedding blocks, BEEM Bojax, 15(e)
Embedding kits,
  epoxy resins, 9, Table A.1
  methacrylates, 13
  polyester resins, 10
Embedding moulds for flat embedding, 17
Epok 533, 9(b)
Epon 812 and Epon 815, Table A.1
Epoxy resins, 9, Table A.1
Encapsulating media, 6
ERL 4206, Table A.1
Ethylene glycol, 7(a)

Falcon Petri dishes, 19(a)
Fibrinogen, 6(b)
Filters, Millipore membrane, 4(e)
    Nucleopore, 19 (f)
Fixatives, 3
Flat embedding, moulds for, 17
Formaldehyde, 3(c)
Freon 12 and 22, 21 (a)

Gelatin capsules, 15(a)
Glass vials, 4(a)
Glycol methacrylate (GMA), 13(b), 14(b)
Glutaraldehyde, 3(b)
Gum acacia, 5(b)

Hardeners, for epoxy resins, 9, Table A.1
Hexahydrophthalic anhydride (HHPA),
    Table A.1
Hydroxy propyl methacrylate (HPMA),
    13(c), 14(c)

Intermediate solvents, 7
Isopentane, 21 (b)

Jurand and Ireland rotary shaker, 8(b)

Luperco, 12 (d)

Maraglas 655, 9(c), Table A.1
Melinex film, 19(c)
Methacrylates, 14
    accelerators for, 12
    embedding kits, 13
    initiators for, 12
Methyl methacrylate, 13(a), 14(a)
Methyl nadic anhydride (MNA), Table A.1
Mica, 19(d)
Microfuge, 4(f)
Micromoulds (sets of plastic capsules), 15(f)
Millipore membrane filters, 4(e)
Mixers, rotary, 8
Moulds,
    for flat embedding, 17
    polyethylene, 17(b)
    silicone rubber, 17(c)

Nadic methyl anhydride (NMA), Table A.1
n-butyl methacrylate, 13(a), 14(a)
Nonenyl succinic anhydride (NSA),
    Table A.1
Nucleopore filters, 19 (f)
Nunclon Petri dishes, 19(b)

Osmium tetroxide, 3(a)

Paraformaldehyde, 3(d)
Petri dishes,

Falcon, 19(a)
Nunclon, 19(b)
Phosphate buffer, 2(a)
Polaron 'peelcap' capsules, 15(d)
Polyester film, Melinex, 19(c)
Polyester resins, 11
    accelerators for, 12
    embedding kits, 10
    initiators for, 12
Polyethylene capsules,
    BEEM, 15(b)
    Taab, 15(c)
Polyethylene glycol 200, 7(b)
Polyethylene moulds, 17(b)
Polypropylene capsules, Taab, 15(c)
Polyvinylpyrrolidone (PVP), 5(c)
Potassium ferricyanide, 3(h)
Potassium permangate, 3(g)
Propane, 21 (c)
Propylene oxide, 7(c)

Razor blades, 4(d)
Resins, epoxy, 9, Table A.1
    polyester, 10, 11
Rhodester 1108, 11(b)
Rigolac 2004 and 70F, 11(c)
Rotary mixers, 8
Rotary shakers, 8

Selectron, 10(a), 11(d)
Shakers, rotary, 8
Stirrer, Zacks teflon, 8(h)
Silicone rubber moulds, 17(c)
Styrene, 7(d)
Substrates for cell monolayers, 19

Taab capsules, 15(c)
t-butyl perbenzoate, 12(g)
Teflon-coated coverslips, 19(e)
Thiokol LP-8, 9(g)
Thrombin, 6(b)
Tweezers, 4(b)

Ultra-violet lamps, 18
UV curing chamber, 18(a)
Uranyl acetate, 3(i)
Uranyl nitrate, 12(h)

Veronal-acetate buffer, 2(c)
Vestopal W, 10(b), 11(e)
Vials, EFFA mini-, 17(a),
    glass, 4(a)
Vinyl cups, 17(d)

Wax, dental, 4(c)

Zacks teflon stirrer, 8(h)

Subject index

Accelerators,
for epoxy resins,
benzyl dimethyl amine (BDMA), 132–135, 138, 140, 141, 144, 157
dimethyl aminoethanol (DMAE), 134, 135, 137
DMP-30, 132–135, 138, 139, 140, 141, 142, 144, 147, 157
storage of, 143
for polyester resins,
cobalt naphthenate, 148–152
Acetaldehyde, 30, 50
Acetone,
as dehydrating agent, 111–117
as intermediate solvent, 111
Acid alcohol, 120
Acrolein (acrylic aldehyde) fixatives, 30, 48–50
with glutaraldehyde, 49, 63
with glutaraldehyde and paraformaldehyde, 50
Acrylic aldehyde (see Acrolein)
Aldehyde fixatives (see also Acrolein, Formaldehyde, Glutaraldehyde and Paraformaldehyde), 6, 9, 10, 19, 24, 30, 50
containing digitonin, 64
containing trinitro compounds, 61, 64, 86
followed by osmium tetroxide, 6–7, 10, 11, 12–13, 24, 26, 31–38, 77
penetration of, 44, 48, 65
perfusion with, 15, 31
Anaesthetics,
application of, 80, 82
disadvantages of, 80
Anhydride/epoxy ratio, 132, 133, 141–142

Aquon, water-soluble epoxy resin, 156, 157
Araldite CY 212 (M), epoxy resin, 131, 132, 133, 135, 136, 140–141, 145
Araldite 502, epoxy resin, 131, 132, 133, 135, 136, 141, 142, 143, 146, 149, 180
Araldite 506, epoxy resin, 131, 132, 133, 135, 140
Araldite 6005, epoxy resin, 131, 132, 133, 135, 141
Azo-bis-iso-butyronitrile (AIB), 153

Bacteria, fixation of, 30, 65, 94, 103–105
Ballistic technique for rapid specimen cooling, 179, 182
Barium permanganate fixative, 57
BEEM capsules, 127, 129, 166, 170
Beetle 4116 and 4134, polyester resins, 149, 150
Benzoin, initiator, 149, 150, 151, 152, 153
Benzoyl peroxide, initiator, 148, 149, 150, 152, 153, 155
Benzoyl peroxide paste (Luperco), initiator, 149, 153, 160
Benzyl dimethyl amine (BDMA), accelerator, 132–135, 138, 140, 141, 144, 157
Biopsies, fixation of, 85–86
Botanical specimens, fixation of, 48, 50, 51, 86–89, 105
Buffers (see also particular buffers),
cacodylate, 8, 16, 17, 20
collidine, 8, 17, 21–23
phosphate, 8, 12–16, 17, 18
Sörensen's, 12–14
toxicity of, 8
veronal-acetate, 8, 18–21

Butanox, initiator, 149

Cacodylate buffer, 8, 16, 17, 20
  for glutaraldehyde fixatives, 41, 43
  for glutaraldehyde–osmium tetroxide fixatives, 58–60
  for osmium tetroxide fixatives, 27–28
  for paraformaldehyde fixatives, 16, 17,
  20, 45–46
  preparation of, 16
Calcium ions in fixatives,
  beneficial effects of, 25, 37, 41
  precipitation caused by, 12, 25, 26
Calcium permanganate fixative, 57
Capsules,
  BEEM, 127, 129, 166, 170
  gelatin, 125, 127–129, 147, 166
  polyethylene, 125, 127, 147, 170
  polypropylene, 128
  TAAB, 127–128
Cardolite NC–513, reactive flexibilizer,
  131–135, 137, 138, 140, 142, 157, 158
Catalysts (*see* Initiators)
Caulfield's osmium tetroxide fixative, 29
Cells,
  cultured, fixation of, 90–91
  isolated,
    encapsulating methods for, 95–99
    fixation of, 7, 9, 25, 31, 60, 75, 91–99,
    105
Cholesterol, preservation of, 64, 102
Chrome–osmium fixative (Dalton), 57–58
Cobalt naphthenate, accelerator, 148–152
Collidine buffer, 8, 17, 21–23
  for glutaraldehyde fixatives, 22, 37, 41, 43
  for glutaraldehyde–osmium tetroxide fixatives, 58
  for osmium tetroxide fixatives, 21–23, 30
  for paraformaldehyde fixatives, 17, 22–23,
  45, 46–47, 86
  preparation of, 23
Colloid osmotic pressure, 8, 9, 84
Cross-linked methacrylate, 3, 154
Crotonaldehyde, 30, 50
Cryo-protective agents, 177–179
  dimethyl sulphoxide, 177, 179, 181
  ethylene glycol, 177, 179, 181, 182
  glycerol, 177, 180
Cryo-ultramicrotomy, 184–185

Dalton's chrome–osmium fixative, 57–58
DDSA (*see* Dodecenyl succinic anhydride)

Dehydration, 111–120
  dimensional changes during, 3, 112
  effects of, 111–112, 114–115
  extraction during, 21, 24, 37, 51, 111–112,
  116–118
  low-temperature, 177–178
  partial, 116–117
  rapid, 116
  schedules, 112–117
Dehydrating agents,
  acetone, 111–117
  acid alcohol, 120
  ethanol, 111–117
  ethylene glycol, 118–119, 161
  polyethylene glycol (Carbowax), 119
  propylene oxide, 119–120
  water-soluble embedding media, 156–161
DER 332, epoxy resin, 134–135, 138, 139,
  146
DER 334, epoxy resin, 131, 134–135, 136,
  142, 143, 146
DER 732, reactive epoxy flexibilizer, 131,
  133, 134, 135, 137, 138, 139
DER 736, reactive epoxy flexibilizer, 131,
  133, 134, 135, 137,
Dibutyl phthalate, plasticizer, 132–135, 137,
  138, 140, 141, 157
Digitonin in aldehyde fixatives, 64, 102
Dimensional changes,
  during dehydration, 3, 112
  duration fixation, 3, 25
  during polymerization of embedding
  media, 3, 125
Dimethyl aminoethanol (DMAE), accelerator, 134, 135, 137
Dimethyl sulphoxide, cryo-protective agent,
  177, 179, 181
Divalent ions in fixatives,
  precipitation of proteins by, 12, 38
  preservation of lipids by, 12, 24–25, 37,
  44, 57
Divinyl benzene, cross-linking agent, 154,
  155, 156
DMP-30, accelerator, 132–135, 138, 139,
  140, 141, 142, 144, 147, 157
Dodecenyl succinic anhydride (DDSA),
  hardener, 131–135, 136, 138, 139–140,
  141, 142, 157, 158
Double fixation,
  advantages of, 31
  illustrations of, 11, 13, 32–33, 39
  need for, 38, 47, 48

precipitation during, 12–13
Durcupan, water-soluble epoxy resin,
156–159, 180

Embedding media,
Aquon, water-soluble epoxy resin, 156,
157
Araldite, epoxy resin (*see* Araldite)
Beetle, polyester resin, 149, 150
containing water, 161
coverslips made of, 168–169
cross-linked methacrylate, 3, 154
DER 332, epoxy resin, 134–135, 138,
139, 146
DER 334, epoxy resin, 131, 134–135, 136,
142, 143, 146
Durcupan, water-soluble epoxy resin,
156–159, 180
effects of, 3, 125
Epikote, epoxy resin, 131, 135
Epok, epoxy resin, 131, 134, 135, 146
Epon, epoxy resin (*see* Epon)
epoxy resins, 130–148
ERL 4206, epoxy resin (*see* ERL 4206)
gelatin, 161–162, 185
glycol methacrylate (GMA), water-
soluble, 159–160, 178, 181
hydroxypropyl methacrylate (HPMA),
water-soluble, 159, 160–161, 166, 181
Maraglas, epoxy resin, (*see* Maraglas)
methacrylates, 153–156
polyampholytes, 162–163
polyester resins, 148–153
preparation of, 126, 128, 143–144,
150–153, 154–155
properties of, 123
protein-aldehyde, 162
Rhodester, polyester resin, 149, 150
Rigolac, polyester resin, 149, 150,
152–153
Selectron, polyester resin, 150
urea-aldehyde, 162
Vestopal W, polyester resin, 148, 149,
150, 151–152
viscosity of, 136–137
water-soluble, 156–161
Embedding methods,
flats, 163–164, 167
for monolayers of cells, 164–170
for small quantities, 170
standard, 125–130
Embedding schedules,

at room temperature, 147, 150
for epoxy resins, 144–148
for methacrylates, 155–156
for polyester resins, 151–153
rapid, 148, 152
Encapsulating methods,
with agar, 95–96, 98
with albumen, 98–99
with bovine serum albumin (BSA), 97–98
with fibrin, 96–97
Epikote, epoxy resin, 131, 135
Epok, epoxy resin, 131, 134, 135, 146
Epon, epoxy resin, 116–117, 119–120,
131–135, 136, 139, 140–141, 142, 145,
146, 147, 148, 180–181
Epon–Araldite mixtures, 140–141
Epoxy resin embedding media, 130–148
preparation of, 143-144
Epoxy resins (*see* Aquon, Araldite, DER
332, DER 334, Durcupan, Epikote, Epok,
Epon, ERL, Maraglas)
ERL 4206, epoxy resin (Spurr), 131, 134,
135, 136–137, 143, 146, 147
Ethanol, dehydrating agent, 111–117
Ethylene glycol,
as cryo-protective agent, 177, 179, 181,
182
dehydration with, 118–119, 161
Extraction of cellular components,
during dehydration, 21, 24, 37, 111–112
during fixation, 6, 10–12, 21, 50

Fixation,
aims of, 5–6,
artefacts caused by, 5, 10–12, 37, 64
criteria of good, 2
dimensional changes during, 3, 25
equipment for, 75, 79
in vivo, 79–80
methods of,
by immersion, 73, 78–79
by injection, 80–81
by perfusion, 74, 80–85
of bacteria, 30, 65, 94, 103–105
of blood cells, 16, 17, 18, 20, 22, 92, 93
of botanical specimens, 6, 48, 50–51,
86–89, 105
of cell fractions, 98, 99–100
of cells with thick walls, 102–103
of glycogen, 36–37, 44, 51, 61, 65
of isolated cells, 7, 9, 25, 31, 60, 75, 91–99,
105

of large specimens, 85–86
of lipids, 7, 12, 24, 25, 37, 38, 44, 48, 51, 58, 60–61, 64, 65, 102, 105
of marine organisms, 15, 104, 105
of membranes, 6, 7, 9, 17, 22, 25, 31, 32–34, 39, 44, 50, 57, 61, 62–63, 64, 65, 67
of microtubules, 10, 48, 76
of mitochondria, 17, 23, 36, 37, 39, 52, 74, 76, 98, 99
of monolayers of cells, 11, 59–60, 90–91
of muscle fibres, 12, 31, 34, 35, 50, 74
of nuclear material (DNA), 9, 17, 24, 51
of organ cultures, 86
of pellets, 91–93
of proteins, 24–25, 31, 36, 44, 50–51, 58
of small quantities of material, 100–102
of spindle fibres, 9, 25
precipitation during, 12–13, 25, 26
primary, 75–76, 78–79
schedules, 77, 104–105
secondary, 6–7, 10, 12–13, 24, 31, 38, 77
temperature of, 75–76, 77, 80, 81, 84, 87, 89, 93
time of, 75–77
Fixatives,
acrolein, 30, 48–50
aldehyde (see Aldehyde fixatives)
choice of, 5–6, 9, 41, 77
containing alcian blue, 65
containing digitonin, 64
containing divalent cations, 12, 16, 24–25, 37, 41, 44, 47
containing potassium ferricyanide, 60–61, 63
containing hydrogen peroxide, 65
containing ruthenium red, 65
containing trinitro compounds, 61, 64, 86
development of, 6–7
formaldehyde, 7, 9, 10, 30, 44–45
for perfusion, 8, 12, 15, 31, 83–85
glutaraldehyde (see Glutaraldehyde fixatives)
ionic contitution of, 7, 8, 10–12, 23
mixed, 57–65
osmium tetroxide (see Osmium tetroxide fixatives)
osmolarity of, 7, 8, 9–10, 25
paraformaldehyde (see Paraformaldehyde fixatives)
paraformaldehyde–glutaraldehyde (Karnovsky), 7, 31, 44, 47–48, 61, 77
penetration of, 7, 23–24, 44, 46–48, 58,

61, 73, 75–76, 79, 85–86, 102–103
pH of, 7, 8–9
potassium permanganate, 6, 50–53
unbuffered, 8, 19
uranyl acetate, 7, 19, 60, 65, 67, 104
vapours of, 7, 90–91, 100
vehicles for, 7–23
Flat embedding,
methods of, 163–164, 167
moulds for 163–164
Flexibilizers for epoxy resins (see Cardolite NC–513, DER 732, DER 736, Thiokol LP–8)
Formaldehyde fixatives, 7, 10, 30, 44–45
chemical and morphological effects of, 9, 44
perfusion with, 15
Freeze-drying, 178
Freeze-substitution, 178–183

Gelatin,
capsules, 125, 127–129, 147, 166
for embedding, 161–162, 185
Glutaraldehyde,
fixatives, 6, 9, 10, 30–43
cacodylate buffered, 41, 43
chemical and morphological effects of, 9, 11, 31–38
collidine buffered, 22, 37, 41, 43
osmolarity of, 10, 38, 41
osmotic effects of, 38
perfusion with, 15, 85
pH of, 41
phosphate buffered, 41, 42–43
precipitation following, 12–13
preparation of, 41–43
storage of, 41
with acrolein and paraformaldehyde, 50
with hydrogen peroxide, 65–66
with osmium tetroxide, 58–60, 76
with paraformaldehyde (Karnovsky), 7, 31, 44, 47–48, 61, 77, 78
solutions of, 38–40
polymerization of, 38
purification of, 40
storage of, 38–40
vacuum distillation of, 40
Glutaraldehyde–osmium tetroxide fixative, 58–60
Glycerol, cryo-protective agent, 177, 180
Glycogen, fixation of, 36–37, 44, 51, 61, 65

Glycol methacrylate (GMA), water-soluble, 159–160, 178, 181
Glyoxal, 30

Hardeners for epoxy resins (see, DDSA, HHPA, MNA, NMA, NSA)
Hexahydrophthalic anhydride (HHPA), hardener, 132, 133, 135, 139, 143, 157
Hexylene glycol, 120
HHPA (see Hexahydrophthalic anhydride)
Hydroxypropyl methacrylate (HPMA), water-soluble, 159, 160–161, 181
  as an intermediate solvent, 166

Immersion fixation, 73, 78–79
  equipment for, 75, 79
Initiators (catalysts) for polyester resins and methacrylates (see Benzoin, Benzoyl peroxide, Benzoyl peroxide pasts (Luperco), Butanox, and Tertiary butyl perbenzoate)
Intermediate solvents,
  acetone, 111
  hydroxypropyl methacrylate (HPMA), 166
  methyl methacrylate, 151
  propylene oxide, 111, 113, 116, 117, 118, 165, 166
  styrene, 111, 117
  toluene, 111, 113, 118
  xylene, 111, 113
Inverted capsule technique, 166
In vivo fixation, 79–80

Karnovsky's fixative (see Paraformaldehyde–glutaraldehyde fixatives)

Lanthanum permanganate fixatives, 55–57
Lipids,
  extraction of,
    during dehydration, 24, 37, 51, 111–112, 116–118
    during embedding, 125
  preservation of,
    by divalent ions, 12, 24–25, 37, 44, 57
    by hexylene glycol, 120
    by potassium ferricyanide, 60–61
    during storage, 113
Low-temperature methods, 177–185
Luperco, benzoyl peroxide paste, initiator, 149, 153, 160

Maraglas 655, epoxy resin, 131, 134, 135, 136, 142, 146, 147, 181
Marine organisms, fixation of, 15, 104, 105
Melinex, polyester film, 165–166
Methacrolein, 30
Methacrylates,
  n-butyl, 153, 154, 155, 159, 160
  cross-linked, 3, 154
  distillation of, 154
  glycol (GMA), 159–160, 178, 181
  hydroxypropyl (HPMA), 159, 160–161, 166, 181
  methyl, 151, 153, 154, 155
  mixture with polyester resin, 154
  mixture with styrene, 154, 155, 156, 159
  partially polymerized, 153, 155
  water-soluble, 159–161
  uranyl nitrate added to, 153
Methyl methacrylate, 151, 153, 154, 155
  as an intermediate solvent, 151
Methyl nadic anhydride (MNA), hardener, 132–136, 139–140, 141, 142, 147, 157, 158
Mica substrate for cell cultures, 165, 167
Microtubules, loss of,
  at low temperatures, 76
  in acrolein, 48
Millipore membrane filters,
  as a substrate for cell cultures, 90, 167–168
  for collecting cells, 99
  for collecting subcellular particles, 100
Mitochondria, fixation of, 17, 23, 36, 37, 39, 52, 74, 76, 98, 99
Mixed fixatives, 57–65
  glutaraldehyde–osmium tetroxide, 164–170
  osmium tetroxide–potassium dichromate (Dalton), 57–58
MNA (see Methyl nadic anhydride)
Monolayers of cells,
  embedding of, 164–170
  fixation of, 11, 59–60, 90–91
Myelinic figures,
  formation of, 36, 37
  prevention of, 41, 47, 64

Nadic methyl anhydride (NMA), (see Methyl nadic anhydride, MNA)
n-Butyl methacrylate, 153, 154, 155, 159, 160
Nonenyl succinic anhydride (NSA), hardener, 133, 134, 135, 136, 137

Organ cultures, fixation of, 86
Organs, fixation of, 78–85
Osmium tetroxide,
    fixatives, 6, 10, 23–30
    after aldehyde fixation, 6–7, 10, 11,
        12–13, 24, 26, 31–38, 77
    buffers for, 23–24
    cacodylate buffered, 16, 27–28
    Caulfield's, 29
    chemical and morphological effects of,
        9, 11, 23–25
    collidine buffered, 21, 23, 30
    Palade's, 28, 57
    penetration of, 23–24, 58
    perfusion with, 15
    phosphate buffered, 8, 12–13, 15,
        26–27
    Ryter–Kellenberger, 30, 65, 103–104
    unbuffered, 8, 19
    veronal–acetate buffered, 18, 19,
        28–30
    with glutaraldehyde, 58–60
    with potassium dichromate (Dalton),
        57–58
    with potassium ferricyanide, 60–61, 63
    with ruthenium red, 65
    Zetterqvist's, 29
    solutions,
        preparation of, 25–26
        recovery of, 26
        storage of, 26
Osmolarity of fixatives, 7, 8, 9–10, 25
    measurement of, 9

Palade's osmium tetroxide fixative, 28, 57
Paraformaldehyde fixatives, 7, 44
    buffers for, 16, 17, 18, 20, 22, 45
    cacodylate buffered, 16, 17, 20, 45–46
    collidine buffered, 17, 22, 23, 45, 46–47,
        86
    for large tissue blocks, 46–47
    phosphate buffered, 17, 18, 45–46
    preparation of, 45–46
    properties of, 45
    storage of, 45
    with acrolein and glutaraldehyde, 50
    with glutaraldehyde (Karnovsky), 7, 31,
        44, 47–48, 61, 77, 78
Perfusion fixation, 8, 12, 15, 31, 74, 80–85
Permanganate fixatives, 50–58
    destruction of cell components by, 5,
        50–51

for plant tissues, 6, 51
    lanthanum, 55–57
    potassium, 6, 50–53, 67
    sodium, 53–55
Phosphate buffers, 8, 12–16
    for glutaraldehyde fixatives, 41, 42–43
    for osmium tetroxide fixatives, 8, 12–13,
        15, 26–27
    for paraformaldehyde fixatives, 17–18,
        45–46
    osmolarity of, 12–16
    preparation of, 14–16
    Sörensen's, 12–14
Plasticizers for epoxy resins, 137–138
    dibutyl phthalate, 132, 135, 137, 138,
        140, 141, 157
    polyethylene glycol 200 (Carbowax 200),
        134, 135
Polyampholyte embedding medium, 162–
    163
Polyethylene glycol 200 (Carbowax 200),
    as plasticizer, 134, 135
    as dehydrating agent, 119
Polyester resins,
    Beetle, 149, 150
    embedding media, 148–153
    polymerization of, 150
    Rhodester, 149, 150
    Rigolac, 149, 150, 152–153
    Selectron, 150
    Vestopal W, 148, 149, 150, 151–152
    Vinox, 148
Post-fixation (*see* Secondary fixation)
Post-mortem changes, 73–74
Potassium ferricyanide in fixatives, 60–61,
    63
Potassium permanganate fixatives, 6, 50–53,
    67
    preparation of, 51–53
    properties of, 50–51
    storage of, 53
Primary fixation, 75–76, 78–79
Propylene oxide,
    as dehydrating agent, 119–120
    as intermediate solvent, 111, 113, 116,
        117, 118, 165, 166
Protein–aldehyde embedding medium, 162
Proteins,
    fixation of, 24–25, 31, 36, 44, 50–51, 58
    model studies with 8, 10
Pyruvic aldehyde, 30

Rapid dehydration, 116
Rapid embedding, 148, 152
Rate of penetration of fixatives, 7, 10, 23–24, 44, 46–48, 58, 61, 73, 75–76, 79, 85–86, 102–103
Rhodester 1108, polyester resin, 149, 150
Ribosomes, extraction by permanganates, 50, 52, 53
Rigolac, polyester resin, 149, 150, 152–153
  mixture with styrene, 149, 150, 153
Rotary shakers and mixers, 126–128
Ruthenium red in fixatives, 65
Ryter–Kellenberger osmium tetroxide fixative, 30, 65, 103–104

Secondary fixation, 6–7, 10, 12–13, 24, 31, 38, 77
Selectron, polyester resin, 150
Sodium permanganate fixatives, 53–55
  for amphibian tissues, 54–55
Softening temperature of blocks, 142–143
Sörensen's phosphate buffer, 12–14
Styrene,
  as an intermediate solvent, 111, 117
  mixture with methacrylates, 154, 155, 156, 159
  mixture with Rigolac, 149, 150, 153
Subcellular fractions, processing of, 99–100
Substrates for cell monolayers, 164, 165, 167–170
Sucrose,
  use of to adjust osmolarity of fixatives, 9, 13, 14, 25, 27, 29, 42 43, 46, 48, 49, 76, 81, 88, 99
  effect on fixation, 25, 36, 99

Taab capsules, 127–128
Temperature of fixation, 75–76, 77, 80, 81, 84, 87, 89, 93
Tertiary butyl perbenzoate, initiator, 148, 149
Thiokol LP-8, reactive flexibilizer, 133, 135, 137, 147, 157, 158

Triallyl cyanurate (TAC), 132, 134, 135, 142–143, 146, 149, 150
Trinitro compounds in fixatives, 61, 64, 86

Ultra-violet polymerization, 128–130, 145, 147, 149, 150, 155–156, 159, 166, 177, 179, 180–181
Uranyl acetate fixatives, 7, 19, 60, 65, 67, 77, 104
  veronal-acetate buffered, 19, 65
Uranyl nitrate for methacrylate embedding, 153
Urea–aldehyde embedding medium, 162

Vacuum distillation of glutaraldehyde, 40
Vehicles for fixatives, 7–23
Veronal–acetate buffers, 8, 18–21
  for osmium tetroxide fixatives, 18, 19, 28–30
  for uranyl acetate fixatives, 19, 65
  preparation of, 19–21
Vestopal W, polyester resin, 148, 149, 150, 151–152
  storage of, 150–151
Vinox K3, polyester resin, 148
Viscosity of epoxy resins, 136–137

Water-soluble embedding media,
  Aquon, epoxy resin, 156, 157
  Durcupan, epoxy resin, 156–159, 180
  gelatin, 161–162, 185
  glycol methacrylate (GMA), 159–160, 178, 181
  hydroxypropyl methacrylate (HPMA), 159, 160–161, 181
  protein–aldehyde, 162
  urea–aldehyde, 162
Weight per epoxide (WPE), 142

Xylene, intermediate solvent, 111, 113

Zetterqvist's osmium tetroxide fixative, 29
Zinc permanaganate fixative, 57

Printed and bound by CPI Group (UK) Ltd, Croydon, CR0 4YY

03/10/2024

01040427-0003